3D Printing Unleashed
7 Key Questions Answered Inside

3D Printing Unleashed
7 Key Questions Answered Inside

Peter Goodwin

Artbot Ltd.

ArtBot

2015

First Printing: 2015

ISBN **978-0-9934958-1-6**

Published by Artbot Ltd.

International House

24 Holborn Viaduct

CITY OF LONDON

London EC1A 2BN

United Kingdom

www.artbot.co.uk

Dedication

To Margaret, for everything.

Contents

Acknowledgements

I would like to thank my family and friends, especially Max Eames, who have helped and encouraged in the making of this book. Without your help my path would have been a stony one.

Preface

Experiencing 3D Printing

What is the experience of having a 3D printer at your fingertips? Imagine having such a potential under your thumb: You control the time, you control the space, and you control the form of the object to be created.

At your command is your very own ever obedient 3D printer. This minion will do what it is told. It will follow your every instruction.

You have conceived your design. From its inception in your thoughts through to the modelling of your idea on a computer. By adding primitive shapes together in a virtual world you start to bring its form together. It shape formed at whim or by cunning calculation.

Behold your desire, your design, floating before your eyes. A multifaceted thing which owes its existence as the product of your effort.

You can see it on the screen in front of you. It is tantalisingly close but as yet without physical substance. It is a whisper of what is to come to reality.

When all is set up and ready to you press the print button of your minion. Your 3D printer appears to do nothing in response. Nothing just yet that is. It is warming to its task. Fans come on and the printer stirs into life, with a cheerful song its motors churn and dance to the tune of a warm up calibration.

Your print design is about to come into being. More parts move and the nozzle crawls across the print bed. Just the silhouette of your design is outlined at first.

As the print progresses you watch your shape start to form. Layer upon layer is added. Each layer is placed carefully on top of another to the accompaniment of the printers own tune: A mechanical cacophony echoing in sympathy to the path it follows. The silhouette is solid now, several layers deep. The magic of 3D printing has started.

Your dutiful minion starts to fill the void within the silhouette with a mesh of honeycomb. Its tempo picks up as each layer takes less time to build as the nozzle moves deftly on. You notice that each layer marks a ridge at the edge of the print. The print head weaves and darts and suddenly two parts of the print become joined as one. You are entranced to see the print growing before your eyes.

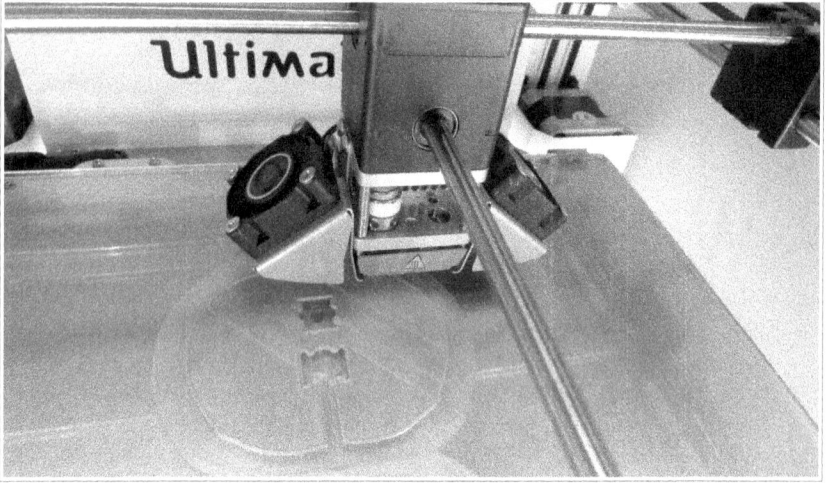

Illustration 1: A print coming to life - courtesy of an Ultimaker 2

How did this miracle of engineering come about? What magic makes it work and how can you take advantage of it?

You itch to know. I itch to tell you.

Introduction – The Exciting World of 3D Printing

Humans often want to create and build in some way or another. Many of us started and have continued doing this since we were small children. Some of us build for a living, others as hobbies to learn and grow. With the act of design and creation we can unlock the innate creativity in us. What we learn from the act of making can enrich our lives and those around us.3D printing[1] empowers us to take our creative impulses further, to hone new skills and ideas and share these with others from all across the world. What we make and create can be shared amongst and enhanced by a global community.

Fuelled by the power of computers and the Internet, 3D printing is a continually evolving technology. It is spanning new horizons in terms of the materials we can print with, the ease with which we do so, and the scale in which we can operate. The art and craft of making is brought to new heights thanks to the exciting world of 3D printing.

This book leads you through the intricacies of this new world. It gives you a grounding in the origins of 3D printing, how it works and highlights the directions in which it may grow in the future.

Whether you are a DIY enthusiast, an artist, programmer, engineer, scientist, fashion designer, or simply someone with the curiosity and drive to take the interest to start reading, this book is for you. I hope it helps you gain insight into, and the courage to adopt, this technology.

Illustration 2: A suitably big printer made by BigRep on show at the 2015 3DPrintshow

Illustration 3: Printing with seeds courtesy of the Print Green Project as seen at the 2014 3DPrintshow

Chapter 1: What is 3D Printing?

A definition of 3D printing

3D printing is a way of making physical objects under computer control by combining and bonding feedstock material together. It is a form of additive manufacturing[2] used to generate an accurately formed three dimensional object. This is done in such a precise way that the process can be repeated to give almost exactly the same result time and time again.

The definition of 3D printing contrasts with other forms of traditional subtractive manufacturing[3], such as machining[4] (e.g. cutting, grinding, sawing, turning and drilling). It can be used to compliment or as an alternative to these traditional methods.[5]

If this sounds a bit technical it is, but the principles behind it are as simple as building with bricks. The next chapter goes into the details of how it works.

What is the attraction of 3D printing

When I talk to people about 3D printing their eyes often light up with excitement. They have read the newspaper articles, seen the news programmes or online blogs. With all this it seems that 3D printing the panacea for creating everything all from your desktop, or the moon.

I showed a young child my 3D printer in action. In awe and fascination he said "Hey, I could print all the Lego I want!"

Some of my other friends say to me "Wow, that would be really great to have. I could print a replacement knob for the cooker that's broken"

3D printing interests different people in different ways. How it will appeal to you will vary and will evolve as you get to know the technology better. It is a tool like many others, but there are certain features and attributes that make it stand out.

It allows iterative change to ones designs and aims. You can take a design, print it and observe it in use. Maybe it will work better being a bit long here, or a handle added over there will make it easier to hold or transport. The cost of change or redesigning something is minimal, just a little time and your idea of something thing is tweaked and the object printed refined.

You do not have to wait for an order to placed, invoiced, checked, packaged, despatched and delivered. It can all happen on your desk.

If you want another print, you just press print again and your revised objects created before you. You still have to wait, but fewer people and processes are involved..

Anyone who loves jigsaws will appreciate the satisfaction you can get when two parts you've part slide together and mate cleanly. Parts can be printed to within a fraction of a millimetre and can even be split or twinned with ease, mirrored or flipped upside down if the mood takes you.

You can collaborate with others easily, just upload your design and decide to share it, and others can derive new objects, sometimes combining designs from several different people and melding them together, or modifying them in ways you hadn't thought of, from all across the globe.

There is a freedom to this and immediacy that unfetters you thoughts. The possibilities feel endless with your only limit your imagination.

Those of us who have been to a Makerfaire[6] will recognise the boundless enthusiasm of other people to create things, whether its craft items, artwork, clothing or electronic gadgets. Having A 3D printer gives you the ability to make a phenomenally large number of different objects using the same process. All that changes is the set of instructions sent to the printer that make all the difference between printing a Doll's head, a model of the Eiffel Tower[7] to producing key parts needed to make a Quadropter[8].

The same level of enthusiasm can be seen at the 3D Printshows[9], with folk in awe as machines spin object seemingly out of thin air. The number of different brands and designs of 3D Printer available is growing phenomenally fast, there's sure to be one that will catch your eye.

My first 3D printer was from a kit[10] for one of the early 3D printers which I slowly and carefully put together. I know every bolt, cable, circuit board, wooden panel, fan, motor and nozzle. I still get an a great sense of pleasure every time I use it, and I still use it to this day. it is been modified and enhanced and works better than it was originally designed to, all with the help of an online community and the help of friends and fellow enthusiasts. You don't have to do this, nowadays you have much more choice and have plenty of different ones to choose from ready built, or even skip owning one altogether and still be able to print in 3D.

If you want to be a part of a community of the enlightened 3D enthusiasts to entrepreneurs, who are prepared to share and collaborate for common goals, find fixes for common problems and develop design ideas, then 3D printing is for you. If you just want to be able to print cool stuff that others have generously shared online, then 3D printing is also for you.

There are certain things you need to know to press forward. Read on to find out what's in this book to help you fire your imagination in a new way.

What we are not going into - How to build a 3D Printer!

The world of 3d printing is moving so fast that if one attempted to write a definitive or encyclopaedia book almost be out of date as soon as it was written. With this in mind this book delivers practical and inspirational advice on how to navigate this quick changing landscape with the potential to revolutionise the way we make things. Some technical jargon will crop up in the course the following chapters but it is not intended to intimidate the reader more enlighten in a friendly and amenable way so you dear reader are not left floundering at the next meeting of enthusiasts or bamboozled by the promotional pitch of your next 3D printer salesman.

Do not expect detailed instructions on how to put a 3D printer together bolt by bolt, screw by screw, or wire by wire. There is information in abundance on how to do this available online for those of us inclined to build such a machine ourselves. This information is constantly being updated by the manufacturers and consumers of the products such that duplicate this information in a book seems utterly pointless. Not all of us, for that matter, will want to get our hands too dirty either and will simply wish to find the most suitable 3-D printer on the market for our needs.

You won't find an all encompassing list of frequently asked questions (FAQ'S) either as a book cannot compete with online search engines to deal with specific problems you may have. There is however, a short section of tips and tricks to give you some guidance on some general issues that are likely to crop up.

What is covered in this book

The aim of this book is to provide clear and succinct information to empower the reader in the world of 3D printing and to encourage participation in that world. It is intended to be an enjoyable and informative ride without an overemphasis on technicalities.

In this book you will find information explaining what 3D printing is, how it works, a brief history of 3D printing, designing for 3D printing, the different materials you can use, the internet and copyright issues, buying a 3D printer, and the future 3D printing. You will find this book a good grounding to give you a compass in the world 3D printing and what to bear in mind when and how you wish to get involved. Additional material is also available from the website 3dprintingunleashednow.com to all readers.

Let's go from thought to reality

Let us go together and see what insights there are to be had in the wonderful, ever-changing and exciting world of 3D printing, that you can be a part of. Realise your goals in understanding the intricacies of the technologies involved and appreciating the innovation and ingenuity with which they are being used right now.

Chapter 2: How 3D Printing Works

How does it work?

It may seem that you need a PHD and a test tube to get to grips with 3d printing. Perhaps the way 3D printing works is a dark art, needing a coven of witches or warlocks to cast dark imprecations to the god of 3D and imbue you with the power to convert thought into matter at a glance. Do you need arcane technical knowledge or spiritual guidance from the other side? Absolutely not. The principles are easy to grasp.

How 3D printing works is very simple. It puts one thing very precisely on top of another and bonds them together in some way. In other words it is an additive process. Something is made by adding material together to leave a useful shape. It sounds absurdly simple: It is. You might say, hang on that's no different from using bricks, what's the big deal?

Bricks are an interesting analogy for understanding the additive element of 3D printing. They are usually moulded in a uniform shape and are bonded together in situ by cement. Sometimes they get put together without cement at all: The Incas of Peru and Central American Mayans[11] did exceptionally well with their dry brick walls. In most cases the bricks were put in place by hand. 3D printing, however, is done mechanically under direct computer control. This means the process can operate with highly precise positioning of the material you use to make a print with, called the feedstock[12]. If you need to build a house with bricks it helps that bricks are a consistent size. Knowing this makes it easier to calculate how many bricks will be needed to build a house. The equivalent of the bricks is the use of consistently sized feedstock. Knowing its feedstock characteristics helps your 3D printer work out how much to print out to make a model house, for example. By using a machine for 3D printing you can print something the same way time and time again. Whether it is worthwhile will depend on its purpose and having an effective design.

3D Printing Unleashed

There is another important aspect to 3D printing. Imagine you're an artist and you want to make a sculpture like Michelangelo's David[13]. If you do it by hand you take a huge block of stone, a chisel and hammer and chop away at the stone to reveal the David lurking within. With time and bandages for the thumbs you could do this again with another block of Stone. You might get another David, but it won't appear identical to the first. There will be subtle differences if you compared the side by side, one thumb too long, one ear too short, perhaps. This does not happen with 3D printing, a second print can be as good as the first, and the third and so forth.

What else is the result of your Stonemasons work? A huge pile of chipped stone, the waste material you had to remove to reveal the David from within that your mind's eye that you wanted to bring into being. Its a *subtractive* process: Removing parts of a material to leave a useful shape.

Being additive a 3D Print doesn't work like this, it only uses its build material (feedstock) where its needed, and you can control how much material it leaves inside. There is no waste, or very little. You can even save on feedstock and the weight of the object by leaving gaps or voids inside the print. If you overdo this you might not have a print that is still strong enough to be useful. The voids can be deliberately left exposed (like Voronoi[14] objects[15]), or hidden inside. To help maintain strength and structural rigidity the software will use a mesh fill pattern instead of just leaving an empty void. You can decide how to set the way in which it does this, it may be referred to as the infill density. You can choose different mesh patterns too, from rectilinear to honeycomb or spiral. A coarse mesh can be set to save on feedstock and reduce the time to make the object. A fine mesh will be stronger and take more time to print. If the mesh is too coarse the print might be too weak for it is purpose, too fine and you waste print material and production time. If you need maximum strength then a one hundred percent infill can be chosen but this is quite unusual.

Like laying bricks 3D printing works is by carefully placing material in a print path. This path often follows the shortest route from one part of the print to another. The layers are bonded together in various ways depending on the type of material used. The finish of the final object will be affected by how this is done. You often see tiny granules, ridges or lines showing up if you look at a print closely. Sometimes additionally finishing is done in various ways to smooth out these printer artefacts but this is not done by the 3D printer itself.

3D printers do not try to print an object all at once. That would be really difficult and there is an easier way to work things. This is done by slicing. When eating an apple you can take a knife and slice it to make it easier to eat. In a similar way you also use slicing to reduce complexity in the process of 3D printing. Suppose you want to make an apple by putting back some slices of apple. You not going to use parts of a real apple but use horizontal slices of a model of an apple instead.

You recreate the apple by putting one slice on top of another again and again until the top of the apple is reached. Each slice of the apple represents a printing layer. So by breaking the design of an object down into slices we can simply the process of building it. Similarly consider making a loaf of bread by gluing individual slices of bread together.

Illustration 4: The 3D print process: From thought, to design, then slicing and printing

By building objects up in layers it is possible to print the inside and outside of an object at the same time. The inside and outside of the object is built up in the same way, at the same time and by the same process. This has significant implications for designing for 3D printing as you can create what would otherwise be separate components all in one go. It makes it possible to print something as intricate as a gearbox as a single print.

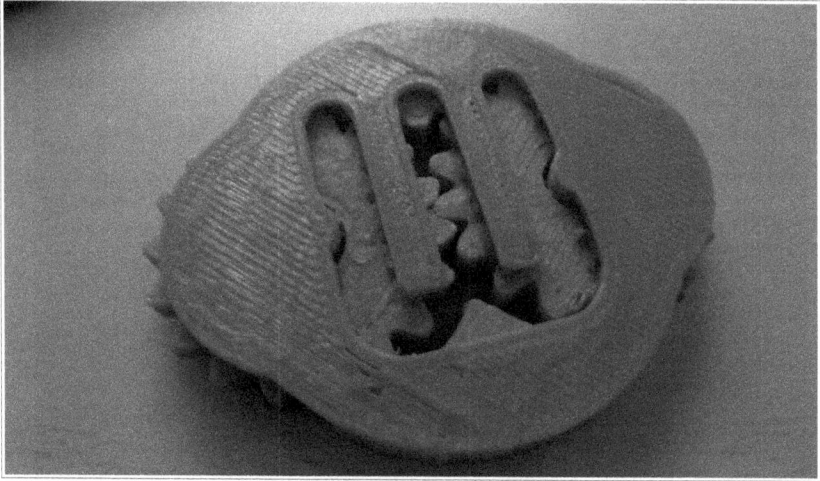

Illustration 5: A gearbox printed in one go - 'Dissolvable Support Gearbox' by tbuser, thing 12342 on Thingiverse

Other technologies require assembling a multitude of parts to achieve this kind of result. With 3D printing you may not even need any screws to put it together, as it already is together. In summary, 3D printing is an *additive* process whereby you add material to combine together to create an object. Most other ways of manufacturing object are *subtractive*, whether by turning, cutting, milling, routing, boring or drilling material is removed to generate the desired object by removing some of it from a stock material.

Slicing

Slicing in 3D printing does a conversion of your model into paths your printer can follow in order to generate a print.

It does not involve destroying an original object to produce it again. The business with the slicing is all done virtually in software. There is no physical slicing required. Understanding the principles of layers and slicing can help in understanding why things can go wrong. For example when orientating your design of a cone it is better that the point of the cone is printed last with the flat end of the cone printed on the platform first to make a stable shape as the print is being produced.

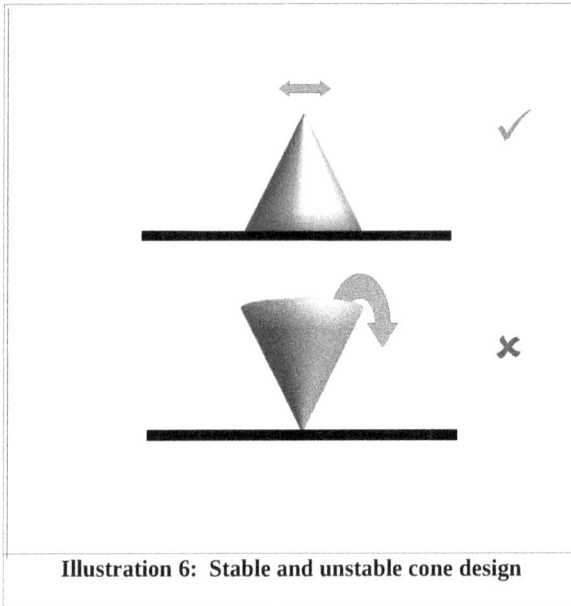

Illustration 6: Stable and unstable cone design

It is possible to print the other way round but this will require support, which depending on the type of printer may affect the finish quality of the print made this way.

There is a trade off in deciding how finely to set the layers. The finer the slicing the higher the quality of the finish, at the cost of a slower print speed. This may be expressed as a quality setting in some software. If you try to set the slicing too fine it may not change the result where the physical limitations of the printer are reached. The purpose of the printed object will have a bearing on this decision too. If you are printing a small hook or clip you might decide on a coarser finish as it is for a utility purpose. For a decorative translucent vase you might wish to get a finer finish as the print will always be under close scrutiny. The size of the final print is also a factor, as a coarse slice will be much less noticeable for a larger print than a small one.

The volume of the print will remain much the same. A fine print will take up the same amount of space as a coarse one. It will need simply need more layers to be printed to make an object of the same height. If you think of the shape of a Ziggurat[16] as opposed to a later Egyptian pyramid[17] you will realise they both describe a pyramidal shape, but the stepped finish of a Ziggurat is the equivalent of coarse slicing compared to the finer smooth finish of the Egyptian design.

Generally the slicing is controlled by the software and printer settings so there is no need to be overly concerned about it, it is simply a factor that affects the final result and finish of the print. See the chapter on software.

What are the main parts of a 3D printer?

A 3D printer can come in a variety of designs and even a brief glance at sales brochures or websites may give you the feeling that it is all too complicated.

The thing to bear in mind is that there are a few main components of any 3D printer. They are as follows:

A chassis. Every printer needs a framework to hold it together and keep it innards working in a precise and predictable manner. It needs to be rigid and strong enough to adsorb the energy of the motion inside so the printer can operate consistently and accurately during printing.

A print head. This can vary in size, shape or weight, from the crossover of two laser beams to a heated metal nozzle. The printer follows a printing path in order to make a print. It is the focus of the printing path that the print head follows to make this work.

A print bed. This is where a print is started off. It can be a chamber where powder or liquid is poured or a flat surface for the first layer to be held in place, ready to bond with the next.

An extruder. This can be a motor or pump used to move the printers feedstock to the print head. It needs to operate to allow a controlled flow which may vary depending on the design being printed.

There a lot so of other ancillary bits of pieces as well, such as the computer electronics used to control the printers movement and communicate with other computers. There are motors, rods, gears and cogs, and sometimes even string required to move the print head about within the chassis correctly. All these parts need to work in concert to give you a viable 3D printer.

Laser based Printing

Laser based 3D printing uses a laser to make a print layer often by triggering a localised phase transition[18] of the feedstock from liquid to solid at a point of focus.

It should not to be confused with laser cutting[19], where a powerful laser is used like a knife to burn away a fine line of a material in order to cut it very precisely.

The way it does this is due to the properties of the print resin, which reacts to the ultraviolet (UV) spectrum of light which is generated by the laser to go solid or start curing. It goes by the name of Stereo lithography and works by having at least two beams of light which meet at a point (intersect) which can be 'moved' or focused to follow a print path and create a layer in the resin. It works this way because the resin only gets enough light to set when the power of the two beams combine: One beam on it is own is not enough to change the state of the resin to change it from liquid to solid. Often the print will need a further stage of curing to be done afterwards to make sure it will set.

This often results in very fine and detailed prints. This is achieved using extremely thin slicing to produce the object. Often the object will need supports, printed as part of the process, as it can suffer from sagging once it's out of the resin pool before it is fully cured and these generally need to be removed manually afterwards.

Extrusion Based Printing

The most common form of 3D printing, certainly if you browse through the serried ranks of manufacturers at a 3D Print shows are based on a process called Fused Deposition Modelling[20] (FDM) which is also known as Fused Filament Fabrication[21] (FFF). The terms Fused Deposition Modelling and Fused Filament Fabrication are interchangeable except that FDM is trademarked by Stratasys[22].

It works a bit like squeezing out a toothpaste tube, only the toothpaste is a polymer plastic (thermoplastic[23]) and the tube is a fine nozzle heated to oven temperatures so the plastic behaves like molten glass and can be forced out as a fine thread from the nozzle.

The nozzle is held in a motorised framework such that it can be moved precisely over a build platform. As the nozzle moves away, the molten plastic deposited cools quite rapidly and turns solid, but it will bond with nearby plastic already deposited or the print bed upon which it has been laid. The nozzle is fed by something called an extruder, which acts to push plastic filament feedstock into the nozzle (where it gets heated) and out onto the print platform. There are many different designs for an extruder, the main variants being direct drive (from a motor spindle), a Wades[24] geared or a Bowden[25] style.

Illustration 7: A Classic Cartesian Printer design with a Cuboid Build Space

Illustration 8: A typical Delta Printer Design with Three Arms

The motorised framework will often fall into two types of design. These are Cartesian or Delta and each have slightly different characteristics. A Cartesian design 3D printer bases its movement in an X/Y fashion, moving a print head (or nozzle assembly) left or right by movement across rods or rails[26]. The delta has the same ability to move its print head, but does this using three arms held like an upside down pyramid with the tip of the pyramid holding the print nozzle or extruder.

In both cases up and down movement (along the Z-axis) is achieved either by moving the build platform (the flat area where the print starts from) up and down or by keeping the platform stable and moving the print head up and down. Different mechanisms are used to achieve the XYZ axis movements do the print head can move left, right, up and down. This defines the area in which a print can be made, called the build volume or build space. This space is not always cuboid, it can be cylindrical as well.

Extrusion printing is quite an efficient system in that although the nozzle is heated to a high temperatures the amount of plastic kept molten at any one time is small, so little energy is need to for the process to work, there is no need for a huge vat or reservoir of molten plastic that needs constantly to kept hot.

One of the downsides of FFF is the potential for weak binding between layers, which can result in delamination or splitting of the final object. Careful tuning of the machines temperature and other settings or new software approaches may mitigate this problem[27].

A lot of the materials[28] used in this type of printing were derived from the injection moulding industry, where objects are produced by forcing molten plastic into a mould, letting it cool then releasing it to repeat the process again. Often used in the mass product of plastic objects it can be scaled up to produce high volumes in one go. Sometimes a 3D printing might have been used to generate the initial patterns for the mould, or it may have been created more conventionally by machining a piece of metal, for example. If you are interested in experimenting with these kinds of techniques on a small scale you could try candle making: A 3D printer can be used to help make the moulds for this too but it is not essential to get going.

Powder based - Binder Jets & Sintering

Do you ever recall playing with sparklers as a child? You know the thin fireworks sticks made of metal compounds that burns fiercely and gives off bright sparks. Some of us may also have played with these and held two of them in one hand together and found that after they burnt they were stuck together. When the sticks burn they generate heat in excess of a thousand degrees centigrade which is enough to meld the metals together when held closely together.

Selective Laser Sintering[29] (SLS) works in a similar way. Sintering[30] is a means of forming a solid mass of metal without fully melting a material using temperature and or pressure. A similar effect can be seen when ice cubes stick together even though they have not been melted to form a solid. A very powerful laser is often used to generate a spot of heat on a bed of powdered metals. The spot of heat is moved in a similar fashion to that of an FDM nozzle point to trace a path so a slice layer can be generated. To print the next layer more powdered metal is added evenly on top and the next print slice is laid out with another scan of the laser. This is often achieved by dropping the print bed by the layer depth and smoothing more print material over the bed with a roller.

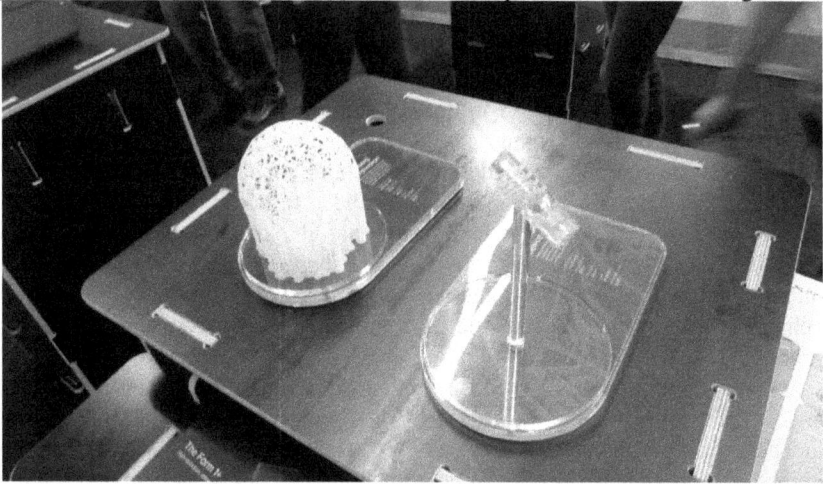

Illustration 9: The wonderfully delicate prints you can achieve with Laser Resin printing exhibited at the UK 2014 3D Printshow

Selective laser melting[31] (SLM) uses a similar technique but in this case the powdered metals fully melted rather than just sintered together. Again very high temperatures are needed to do this. Yet another technique does away with using a laser, and simply sprays a jet glue to adhere metal particles together to form layers. Once all the layers are glued on the print is placed in a sand casting mold, to support the print, while molten metal is poured in to replace the glue and bond the particles together. The process is called Binder Jetting and is handy for metal foundry owners for industrial production runs.

One of the difficulties with these techniques is the tendency powdered metals to burn a bit like a sparkler instead of forming the shape that you want to. This complication to the process implies the need to exclude or reduce the presence of oxygen while the laser is doing its work. However, it is not just metal powders that can be processed in this way but also glass, ceramic, and plastics as well. One of the great advantages of these methods is because the print is in a bed powdered material there is no need for any support structures and one can print designs that might otherwise be fiendishly difficult to print using FFF methods. You do not get any of the stringing effects as with FFF or suffer any warping of the prints either.

The likelihood of a desktop version of these technologies is still some time away. Those printers currently available often cost many thousands of pounds or dollars, may require fume extraction and controlled environments and are unlikely to be suitable for use in domestic environments. This hasn't stopped people trying. Computer Numerically Controlled[32] (CNC) machines have been adapted to this purpose by adding lasers and modifying the bed area to experiment with this. Another approach has been to modify a paste extruder to extrude stiff metal clay, that is a water based clay that has metal flakes embedded in it. Once printed to the required shape, the print is dried then put into a high temperature kiln so that the clay binder evaporates leaving the metal behind as the finished object. The print shrinks somewhat as a result of this process.

Many of the patents for SLS printing[33] are coming to an end recently so expect there to be many more crowd sourced projects attempted in this field in the near future. On the cautious side the technologies are more difficult to implement on the engineering side, much more energy is required in the production so the challenges are steeper then they were when the FDM patents expired, and the costs will be much higher then the FDM equivalents.

Chapter 3: A Potted History of 3D Printing

Older than you might think

From the media, press and television you might have thought quite reasonably that 3D printing is a recent innovation from about 2009, after all how many of us can remember being aware of the term ten years ago? Why has it taken so long to gain popularity and recognition? What are its inherent benefits over other types of technology?

To understand this we need to look at the origins of 3D printing. Computer Aided Design[34] (CAD) software started to change the way people designed products and objects. I well remember being in front of a creaky IBM PC[35] clone in the 1980's running a CAD program which slowly produced the most amazing intricate spider's web of wireframe[36] lattice lines. I was in awe. Nowadays you can get simple apps on your smartphone which exceed the capabilities of what was available at that time. This software was a big change from drafting with paper, pencil and rubber on an easel. It was easy to measure, modify and change a design to fine tune what you were doing. The cost of change was reduced and designs could evolve more quickly. However, the production processes of getting things produced was still involved, time consuming and expensive. The designs need to take into account how products need to be assembled. They needed to ensure that bolts had somewhere to go and to the right depth so that the final product could be put together and stay together, for example. Just think of a petrol motor engine and consider the huge number of parts involved[37].They all must fit properly and allow assembly in the correct sequence for it all to work as a completed unit. For cars clay models[38] were often used to generate prototypes for the final designs, so people could see and check how the designs were going.

This innovation led to pressure on the manufacturing side to provide a faster way to get products prototyped. This became known as Rapid Prototyping[39] (RP) in the 1980's. The most common technique was to use the laser printing I mentioned earlier in the book, but there was also some FDM systems and laminated object printing. Some synergies with the techniques of Computer Numerical Control (CNC) devices were also deployed in terms of the controlling mechanisms and computer systems. This earlier technology derived from the 1950's and enabled automated lathes, routers and milling machines for industry to be used to replace manual machinists.

It is somewhat ironic given what we know now that the first patent for what was then know as Rapid Prototyping (RP) was way back in 1980 by a Dr. Kodama in Japan[40], but that the patent was allowed to lapse without being fully completed.

One can attribute the changeover from the concept of rapid prototyping to the reality of 3D printing with the key development of stereolithography apparatus[41] (SLA). Stereolithography is a term coined by Charles (Chuck) Hull[42] in 1986 patent, from an idea he had in 1983 from experience in sealing desks with Ultra Violet (UV) sensitive plastics. Charles Hull went on to co-found one of the main 3D printer companies of the era called 3D Systems Corporation[43]. It was this company which went on to produce the first commercial 3D printer in 1987 called the SLA-1[44].

A similar technology was also being developed around this time, called Selective Laser Sintering (SLS) by Carl Deckard in 1987.This differed sufficiently to have a patent issued in 1989, and was to end up being acquired later by a company called 3D Systems.

Also in 1989 Scott Crump filed a patent for Fused Deposition Modelling (FDM) which was issued to another company called Stratasys[45] in 1992. A final part of the main 3D printing jigsaw was also with a laser sintering process pioneered by the German EOS GmbH[46] who started selling their Stereos printers in 1990. Although further innovations and techniques have been developed and patented since these are are the main strands of 3D printing technologies that dominated the 3D Printing landscape up until 2009.[47]

So the stage was set for domination by patent owners for a good twenty years of the core technologies we associate with the current the 3D printing techniques available today.

The Patent Effect: What Held us back

The early pioneers of the industry protected their inventions with patents to protect their intellectual property rights for a period of time after their first invention. During this period they enhanced and improved on the basic design, but generally 3D printing remained corralled as a solution for rapid prototyping with the market clearly focused on the corporate world.

The side-effect of this was that technology could only be exploited and developed by those within the companies concerned, which was small microcosm of the potential talent available externally or indeed globally. Moreover the companies themselves were focused quite reasonably on exploiting their advantage getting a good return while the shelter these patents lasted[48].

So what, we may ask, happened when these patents started expiring? Others were to take interest and exploit and develop the potential to make the technology more readily available, the more people and it is significantly lower cost. This blossoming of activity I call the explosion of interest, which I believe is still going on today and which you yourself are also taking part.

The Explosion of interest

A signifier for the explosion of interest in 3D printing is highlighted in the development in a 2005 project called RepRap[49] which was was founded by Adrian Bower, funded through the University of Bath, and was originally focused on designing a 3D printer that could self replicate: The idea was to design a printer that could print the parts for another duplicate printer of the same design. The term RepRap was derived from a shortening of the terms Replicating Rapid Prototyper. All software relating to the project was released under collaborative friendly open source[50] software licences. The concept behind the project was to allow people to produce and distribute printers cheaply when without the need for complex infrastructure or tools and equipment to produce the printers. It relied upon the Internet to distribute the project files schematics and software itself. Another feature of the project was the way in which the various communities that were built up came up with modified designs to improve and enhance the different printer options that became available. In case you are puzzled by the naming convention of the different printer designs they are all named after famous biologists. The first version was named Darwin and other names include Wallace, Mendel (several varieties), Rostock and Huxley. Rostock is somewhat different in that it is a Delta design as opposed to the Cartesian architectures of the others. The main feature of these fused deposition manufacturing (FDM) printers are the use of stepper motors, timing belts and the use of z-axis lead screws. The electronics is based on the open source Arduino[51] range of options. They are also typically without a full enclosure thus avoiding any potential infringement of the Stratasys patent for this feature[52].

RepRap and the internet effect

One of the interesting features of the Reprap project's heavy reliance upon the Internet not only as a means of cooperation between the different members of the reprap group all from different institutions in different continents, but also in the dissemination of their ideas and designs to a global public.

This meant that not only could interested parties join the discussion about the features of the latest thoughts on design and construction, but also adopt the open source designs modify them. This encouraged design variations to improve upon the original designs or accommodate the requirements locally sourced materials and components needed to build the new reprap designs. The sheer energy and momentum that developed led to a global ecosystem in which to support and sustain a new open source based movement which has resulted in a proliferation of many different printer designs at lower prices within the reach of many more people.

The renaissance of craft production of materials has taken a new turn and incorporated many new 3D printer owners who have started calling themselves makers[53], the creation of open repositories[54] for sharing print designs and idea, social groups dedicated to 3D printing and a hive of different commercial companies started up to support the evolution of the market and to promote and sustain the activity.

Without the internet the same momentum and interest is unlikely to have been gained and certainly not in the same timescale. It may even have remained an isolated project limited to the academic circles in whose origins it began.

Impact of History: Legacy of the past

3D printing was originally considered an engineering niche for rapid prototyping. What with the stifling effects of the patents on technological developments and the high price of the 3D printers themselves is not surprising the development of 3D printing remained somewhat stunted in growth. The small size of the industry also meant that economies of scale were less achievable and many of the technologies lent on the developments in more mainstream fields such as computer numerically controlled (CNC) devices.

The way the print platform or the print 'head' moves derives from earlier developments with CNC machines. This includes the very code, called Gcode, that is used to tell the printer what to do to make a print is an extension of the code also used to tell CNC machines what to do.

This heritage that can be seen in the persistence of the Gcode[55] used to control 3D printers, which has had additional commands bolted on to accommodate the needs of 3D printing. There may well be a good case to replace the Gcode control code with something more up-to-date, possibly a custom operating system written specifically for 3D printing. This might offload the current tendency for designs to be sliced by software running on your pc or tablet to a situation where this is handled by the printer itself. Indeed there is a commercially based movement to do just this called spark[56].

Smarter printers also implies the need to enhance the processing power and capabilities of the micro-controllers used inside many of the current designs.

Another constraint that shows in many of the physical 3D printer designs is that of having a single printing focal point or nozzle. While increasing the complexity of the design there is plenty of potential in having multiple focal points or nozzles operating at the same time and perhaps even working in more than one axis, thus greatly increasing the speed with which 3D prints may be paid. Some experimenters have attempted to do this already, albeit in only in one axis, by using modified jet cartridges to see how far they can get in exploring how this might work.

Where's the money?

Some of you may be thinking this is all very well, but how do you make money from 3D printing? The answer is, in a way, you don't. What do I mean by an answer like that? Well if you take the analogy of the American Goldrush[57] it was the guys selling the shovels to the miners more than the majority of the gold diggers themselves who made the money. So you might say there's more money in selling 3D printers and the associated paraphernalia then making money printing things out. There may well come a time when, like in 2D printing, you will get the local printing bureau in the high street or shopping mall who will take your design and print it out for you while you wait. To make money there will take financial muscle and economies of scale to be competitive. The bureaus may not prove necessary, however, if large numbers of people simply upload their prints online and have the prints shipped out to them instead. Companies and community groups[58] are already starting to do this. For the latest information check out www.3dprintingunleashednow.com.

Chapter 4: A New Paradigm?

A panacea for all manufacturing ills?

If you've be reading some of the articles on 3D printing you will have heard the term '3D Printing Revolution' with the implication that it will transform the way we live our lives. While this may very well come true it is worth bearing in mind that the industry is still in it is early stages. I feel it is reminiscent of the computing industry in the 1980's with large shadowy corporations lurking in the background focusing their interest on the high end niches in the market and watching warily as a myriad of small companies launch enthusiastically into the gaps left.

I remember visiting the early 1980's computer shows and being fascinated by the wide range of companies touting their wares. There was a large number of computers available then too, from the Jupiter Ace, Dragon Data, Rair Microcomputer Corporation to Sinclair Research[59] to name but a few. You haven't heard of them? That's because they died off from competition, their ideas were too ambitious for the time or were eaten up or out competed by the big boys. Some companies were trying to do crazy things with their primitive systems, voice recognition or new types of key entry like the Microwriter[60]. They were stretching the boundaries of what could be done, what would sell and the enthusiasm was infectious. I feel the current state of the 3D printing industry is in a similar phase, where once critical mass in the market is there the big corporations will swoop in on the successful companies and buy in their innovations. The buyout of 3D print manufacturer Makerbot to Stratasys[61] can be seen as a case in point.

So with the resonance to the history of computing in mind there will be products and services available now that won't be viable or stay the course. There's nothing inherently wrong in this, you just need to bear this in mind if you start looking out for 3D printing companies, filament supplies or 3D printing and design bureaus.

The other thing to consider is that 3D printing is not an end in itself. If you have a business idea it is important to keep a sense of perspective in that many other ways of producing objects already exist and may be more appropriate to a particular market.

Let me given you an example. Take the ubiquitous plastic spoon. 3D printers can print plastic so why not print spoons? Well in practical terms its a non-starter. Yes you can print a spoon with a 3D printer and there are designs available online that you can use to do this. Consider this, though, in the time it takes to print one spoon, hundreds or thousands can be made in the same time with injection moulding and currently at a much cheaper cost[62]. However if you want a personalised spoon with your name on it or a fairy stuck on end and you only need half a dozen then it will work out cheaper. It is estimated that a new design for a Lego brick to be ready to roll off the production line is many tens of thousands of dollars[63]: The cost of changing or adding production lines, making new moulds, let alone additional costs to packaging and marketing. Yet you can do it for a fraction of the cost with a small 3D printer, some feed stock and the time to download and print a design or design one yourself. If you had to produce thousands of spoons instead of dozen the economics would work against you, but to do a small run the 3D printer can win. If you need to vary the design before each print then you also win with 3D printing. This is because the instead of the changing the physical injection moulding you only requires changes of the design in software. It therefore follows that 3D printing is most suited to low volume high value products, but not the other way round.

Mass production in your Garage?

There have been some talk of having garage factories and I well remember seeing a TV documentary (An episode of Tomorrows World[64]) many years ago featuring a Japanese gentleman who lifted his garage door to reveal an interior packed to the ceiling with various bits of machinery which formed his production line, merrily buzzing away.

One of the issues with 3D printing is that its relatively slow to make an object, with the bigger the object the longer it takes to make proportionally to fill the volume. However, if you use multiple 3D printers running at the same time you can get round some of this restriction, especially if you've split the design up into separate parts that can be put together. You may have done this to get around build volume restrictions or because you want to print it in parts on multiple printers to speed up the production rate. Some 3D printers have been designed to be stackable, so you can create a 3D printing wall in order to maximise the use of space.

The US military used specially equipped cargo containers to be used to repair military equipment and generate spare parts to save the time to ship and space to store a spare parts inventory. They used these in Afghanistan[65]. The cargo containers included 3D printers to allow them to do this.

The power requirements of a 3D printer can be relatively low, about the same as laptop[66]. If you need a heated bed the power need will be a bit more[67], similar to a desktop PC (See the chapter on a buying a 3D printer later on). So you will not need industrial power connections for your 3D printing farm unless you are really thinking big. That means that mass production in your shed or garage is quite feasible, although you will also need to ensure it's watertight and wind-proof.

Implication for Design decisions

3D printing can have quite a profound effect on design decisions. In the early days of the industry use the technology was restricted to making models to show what designs might look like. These models enhanced the understanding of how an item could be manufactured and a product put together. In time firms found that instead of non-functional models they could make functional ones, especially as the cost of printers and their the print feedstock dropped. They soon started to appreciate the quality obtainable with 3D printing. They found that they could design and produce prints that cannot be readily moulded or forged in the conventional ways but which also were good enough to be used in final production processes[68].

Parts that may have required many thousands of pounds or dollars in tooling costs can be made for a fraction of that cost. In addition to this saving the investment in a 3D printer can be spread across many different types of designs and projects. The cost of changing the design with a 3D printer is the time and money spent on the virtual design as the printer only needs fresh instructions to switch from one purpose to another.

The increasing variety feedstock materials also offers the designer a wider range of materials to use in their designs and their ability and the ability to mix and match materials for a given design purpose means there are more ways in which 3D printing become part of the production processes.

When designing for 3D its is important to realise that your design environment allows you to create objects that might not survive in the real world. The design world you are likely to experience does not take account of real world physics like friction or gravity. You can go ahead and design an object which is almost impossible to print. This means that you can design an object which is not strong enough to sustain itself physically and would collapse under the force of gravity or be too fragile to handle should you manage to print. Of course you can print in the zero gravity environment of space[69] but very few of us have the privilege to experience this. You have to factor that into the design yourself.

It follows that you need to make sure the walls are of sufficient thickness to support the likely loads they will experience in use. So when designing we need to think about how the object will need to function and add our intelligence into the design from our knowledge of the expected use it will get and the environment in which we expect to use it.

If we need to design a waterproof container or a mini submarine then we need to make sure that the skin of the object is watertight otherwise it will sink. Whereas in designing a container you will need to make sure that the tolerances and clearances for the lid of the object are close enough to make a snug fit. It must also be loose enough for the lid to operate cleanly without snagging or catching on the edges.

So there are many design decisions that we can make in our object to ensure that we more successful. The more feedback from others we get on the designs that we upload to online repositories the better our understanding will develop.

Another consideration especially for FDM printers is that we need to account for how it will sit on the print bed. A small narrow foot for the object will probably mean we will need to print with support if the rest of the object is more extensive and this will need to be removed once the objects been printed. If we rotate the object and place it another way will this produce a better print and does the design lend to the flat space surface the print bed will offer? Where a design has moving parts we need to consider what the range of movement will be interact with other parts of the print in a controlled way: There's no point in designing a set of cogs to turn together if there is no support to keep them in place and they would simply fall apart.

Support structures can be added during the slicing process which can compensate for print areas that stick out too far or beyond the ability of the printer to bridge across. A good design factors in issues such as these by minimising overhanging sections, considering the print orientation to be used and ensuring suitable wall thickness's to reduce the risk of a failed print and minimise the need to rely on the printers software to compensate.

How much material needed for a design is worth consideration. 3D printing can be slow at the best of times and it may be that you can hollow out and remove certain parts of your print design in order to minimise the amount of material time taken in order to produce the desired object. The aerospace industry often designs to minimise weight and space as part of their design criteria by creating holes in objects without compromising their structural integrity. You will often find honeycomb patterns and parts produced with rounded holes in them specifically to reduce the amount of material used and yet sustained strength required for the design to meet its intended purpose. We too can adopt some of these principles to improve the quality of design before it ever gets to print.

Don't forget that the material you expect the design printed with will vary its characteristics as well. Some materials will be quite stiff and others more flexible but weaker. If you are printing a particularly large or interlocking object then these factors will become all the more important.

Speed to Market– from weeks to days

Given the early emphasis 3D printing is on rapid prototyping it comes as no surprise to realise that once you can use 3D printing and production potential turnaround from designed product can also be rapid. This ability to respond quickly to changing market demands a new requirements means companies and individuals are able to capitalise on this rapid response to capture market share and become more dominant in the markets in which they do operate.

What this means in practise is that one can design on a rough-in basis, explore what the product might turn out to be like, and iteratively redesign what is needed, as many times as necessary to satisfy the design stakeholders, before committing to high volume production set-up costs or equipment. As a designer it can give you an edge to demonstrate in a physical way the attributes of your design concept without heavy investment costs or long delays in obtaining samples.

The Craft Revival

The power of 3-D printing is now readily accessible at very little cost. This means many people can get their hands on equipment that can produce items very quickly for their own purposes and to share with others. The element design in this process gives scope for personalising or customising print designs to suit individual tastes and aesthetics. If you fancy a coat hook with the emblem of a lion on it, to give a trivial example, you can go ahead and design one for yourself without having to call on the resources of other people to manufacture it for you.

You may even wish to collaborate some design concepts with others online and from communities interested in particular hobbies or purposes. One example of which is the various types of custom quadropter designs which people have uploaded to the online repositories, another is the huge interest expressed making and designing mobile phone cases. Here one may find designs that handful of people are interested in using and yet they can share their experiences and expertise amongst themselves with relative ease. This capability of 3D printing is reminiscent of the arts and crafts movements of the past where a particular taste or design ethos could be followed using readily available tools and materials in the home setting.

Implications for Business

From a mainstream business perspective 3D printing can be considered as simply another black box in the production process, providing additional options and tools to get a job done. It does not mean that a business or entrepreneur has to throw away all the other subtractive ways of making things. A blend of both can lead to a better and cheaper way of operating than focussing entirely on one path or the other. That said there are indications that for some particular business working wholly in the 3D printing field it is possible and profitable.

One of the key advantages of adopting 3D printing is that allows mass customisation of the product. This means if your business is say, producing one-off mugs with customers faces embossed on them, then the conventional means of achieving this would incur fresh tooling costs for each print which would then be discarded, which is not the case with the 3D printing option. However, if one required thousands of mugs with the *same* face embossed on them, say of a pop star, then the economies of scale with conventional production will win out. But then, this is no longer mass customisation but a more conventional product requirement. One could still use 3D printing to produce the prototypes and perhaps the precursors of the production moulds even so.

For small production runs the lack of tooling costs of 3D printing offers a cheaper alternative to conventional options of injection moulding or machining. The film industry has been keen to adopt the technology for producing props and models used in filming. The ability to reprint a particular prop if it gets damaged (accidentally or deliberately) is also a boon given the speed with which 3D printing can achieve for small numbers of intricate items. Moreover if there are changes needed to the props the modifications can be incorporated rapidly and often at little additional cost to the designer.

Business already reaps a huge benefit from the availability of the internet. Not only does this medium allow contact with suppliers and a marketplace for products and ideas, it also allows cross-border collaboration at low cost. For 3D printing this means access to specialist 3d printing supplies and designs for the printers but also the capability to co-operate on design projects across the globe, share design and productions problems and solutions making a matrix organisational design ever more feasible.

3D Printing Unleashed

The adoption of 3D printing can also have a less obvious benefit, in reducing the need for huge inventories of components needed to produce a particular product. Granted with current technology electronics components and circuit boards are not yet feasible with current 3D printers there are signs that with printable conductive 'inks' some elements of capability are approaching in the foreseeable future. The reduction of other components to 3d printer feedstock means that local sourcing of commonly available feedstock reduces the need to transport. A reduction in production line complexity simplifies operational management and allows better focus in obtaining efficiencies of production. Production lines can also be more responsive to changes in of design in turn allowing a more competitive response to external competition or changes in market preferences.

Some questions to think about when considering using 3D printing in your business:

- *Is speed to market critical to maintaining your competitive edge?*
  If yes then incorporating 3d printing in the design process may improve your companies responsiveness to changes in the   market or consumer or stakeholder demands.

- *Does your business involve high levels of production?*
  If so you may need to consider 3D printing in terms of improving design flexibility and responsiveness more than a   solution for the production line itself. If not then you may like to   explore using 3D printing for parts of the production line   process for your product.

- *Do you have to keep high levels of inventory of production components?*
  If yes then 3D printing may help to reduce the number and complexity of your production requirements.

- *Do you have time or resources to explore the potential new technologies offer to improve the way your business is run[70]?*
  If no then maybe come back to this and see in what way your competitors may have adopted 3D printing. Its not always easy to adapt to the changes a new technologies may impose in business and it may take significant commitment to make it work for you. It is worth taking this into account before making key decisions.

Hopefully you will now be in better position to evaluate if 3D printing is right for you and whether to take things further.

Chapter 5 Designing for 3D

Why design for yourself?

Designing something takes time to learn to use the tools, think of an idea, try it out then try it out again and so forth. Why, you may ask, should I bother with all this? You can, after all, download many and varied objects already uploaded and freely available online. Downloading designs is a good way to explore what can done with 3D printing and to enjoy the content that others generously give to the world, bearing in mind that all 3D designs are copyright and restrictions apply to their use (see the chapter later on copyright).

However to limit yourself to the output of others restricts the possibilities of what you can do with a 3D printing. You may also find yourself wanting something and finding there isn't a pre-existing design for it, or not on to your taste.

It is all very well to use the designs of others to meet a particular purpose, and often checking for modifications for your printer proves a fruitful exercise. Yet you will find that a particular need is not catered for or you have had a cracking idea for something while you were in the shower and you just must bring to life. The very idea make you itch to get it done. Yes it might not be very quick and may mean more work that you anticipated but the feeling of watching your printer print your design for the first time is like no other.

Designing for yourself is a gratifying thing in it's own right. The feeling of converting something only you thought of into physical reality is very satisfying. It is also a great feeling to have one of your designs put online to share with others as a global audience. Online repositories often allow you to you see when other people have used your design and benefited from your generosity and ingenuity. The feedback can be encouraging and may lead to an evolution of your design. Other people may take your design, if you permit it, and modify or mix it with other designs and so enhance it in ways that might not have occurred to you on your own. Contributing in this way is a great opportunity to 'pay back' to the printing community for the designs that you yourself have taken advantage of.

Scanning and 3D printing

Is there an alternative to design it all yourself? There's another technology that's often associated with 3D printing: It is called 3D scanning. By copying the outside shape of a physical object you are taking a short cut in a design process that allows you to mix and combine elements of a scan in new ways. It is like making a 3D version of an object silhouette. Bear in mind this does not exclude you from copyright law on the original and can leave you exposed to legal issues you may not expect[71].

3D scanning is a way of taking information about an existing object and storing it for later use. To give the topic justice would require a whole book let alone one small chapter. However, some of the techniques we have already touched on for 3D printing apply in this field too.

The scanning process acts to create a point cloud of data. The data is in the form of the X, Y and Z co-ordinates of usually a large number of specific points that approximate to the external shape of the scanned object. The scanner does not capture information about the internal structure of the object (unless it is a medical scan) and is usually non destructive to the original, depending upon the method used to do the scanning.

Such point cloud data is not intrinsically usable for 3D printing, it needs reprocessing to form polygon or triangular mesh data models, which can then be imported into Computer Aided Design (CAD) software. Quite often the mesh will have so much data in the model it will need to be simplified by various mathematical processes in the editing software to bring it to a more manageable form for 3D printing purposes. That is not to say they cannot be used directly at all, just that the overhead in doing so is best avoided.

There are quite a few companies that provide a service to scan you in 3D and then offer to print a model of yourself as a 3D print, often using ceramic print material and with the option for them to have coloured to match your original scan. These look very effective and are quite delightful to hold, though this may be a fad whose popularity will wane with their growing familiarity.

The processes for 3D scanning vary. There are physical scanning methods, such as sensing an object with a probe using a carriage system or articulated arm, or in some case, a mixture of both. These type of scanners can be very precise and may be able to scan into crevices or some interior areas of an object to gain additional information that would be otherwise harder to obtain. They can also damage the original object as physical contact is required for them to work, so may not be suitable for delicate items or those of high worth.

The type of scanner you are more likely to meet outside a laboratory or factory are the ones that scan by light, either from lasers (coherent light) or using structured light (where light is generated in a structured way, often in grids or bars). Some of these scanners work by using the speed of light as a known constant to calculate how far away a particular point of an object is by timing the delay in obtaining a return reflection, operating like a range finder. With this process you only get a snapshot of how the object appears from one place, and it may need to have several scans taken from different places in order to obtain a useful result. This can be done by moving the scanner round the object, usually in an arc, using mirror deflection to effect a virtual positional change or by using multiple scanners in different locations. Its great for scanning outdoors or for large objects and its relatively quick in scanning. The accuracy of this type of scanning is less than can be achieved by some of the other methods so if you need a closely detailed scan there are better options to choose.

The more common type of scanning you are likely you come across works on the basis of triangulation. The system works by having a laser light source or through use of structured light (that is light that is emitted in either a grid pattern in bars). This light will shine on the object and show up as a line or dot. The position of this within the frame of the scanners sensor helps to work out where its, using triangulation of the distance between the laser emitter and the sensor, the angle of the laser emitter to the sensor and the angle of the returned light form the object. Another variant of this works in a similar fashion, but also relies upon a preset calibration image of patterned dots as a backdrop to the image. The David 3D scanner[72] is an example of this kind of setup and users a planar hand held laser as its light source which you wave slowly over the object being scanned. The accuracy of the triangulation system can be very high, in micrometers, although its scanning range is often limited to within a few meters.

I had the pleasure of being scanned at one of the 3D print shows with a commercial scanner which has multiple sensors mount in an arc on an arm from floor to ceiling. It worked very quickly to produce a print, the slowest part of completing the scan was the processing needed to be done afterwards on a PC. The result was very good except that when it was sent to me some time afterwards the top of my head was missing and so were my legs...I had worn dark corded trousers and the sensors hadn't picked up enough reflections bounced back to get the details of my legs. So even commercial grade scanning can run into unexpected problems.

This is a great technology to use and has many great applications for use in industry, medicine, in preserving cultural artefacts and making it easier to provide copies of ancient objects that might be too vulnerable for people to touch the original, for example. For 3D printing its great to be able to take some old broken part for a product no longer manufactured or supported, and produce an unbroken/repaired replica to replace it and resurrect an item of interest this way.

I would argue, however, that mixing scanning with 3D printing design is not an ideal combination. For one thing you are duplicating something that already exists or the work of another creator. While one can make a point that all designs rest on the shoulders of others, to simply clone what already exists seems a dry and unduly limiting approach with what can be achieved with 3D design. There may also be copyright implications for duplicating some objects, although this unlikely to apply to the figure of some ancient Sumerian war god! The other point is that scanned data does not include any concept of intended or function purpose of a design and has no or little representation of any internal features or structures. Lastly the results of scans are often very messy, with many scatter points that need to be cleaned up to make a useful print, and the resulting meshes lend to having very large numbers of elements derived from the point cloud which adds to the technical difficulties when attempting to manipulate them.

Parametric Designs

A parametric model or design is one that can be changed and resized using value settings within the design specified for this purpose. Not all design software or formats support this, one such that does is Openscad, which is good software to pick to learn about this way of designing. Bear in mind that this may result in distortions or instabilities if the design parameters are set beyond the original creators intention, so a chair design that is originally defined with 5cm legs may result in legs too weak when scaled to metre lengths, but may still be adequate if a smaller scale change is made, say from 5cm to 6cm.

What to consider when designing for 3D Printing

There are a number of factors that can catch you out when designing for 3D printing. Some of these may seem obvious but they can easily catch you out.

Missing Bits

What do I mean by missing bits? Well check for edges that don't quite meet, bad joins or unintended holes which can lead to the dreaded non-manifold errors reported when your design goes for slicing. Some software in use is more focused on rendering 3D designs than fabricating them and may not warn you of these kinds of errors.

Tolerances

There is nothing worse than mixing up your design with elements using different units, millimetres or inches. Be consistent and aware of the origins of your design elements.

Excessive mesh counts

This often occurs when using the results of 3d print scans or the result of conversion processes. A large number of meshes can cause other editing or slicing programs to choke on the input or otherwise fail. There are several programs that can reduce mesh complexities for you so its a good idea to check before uploading to a repository and getting complaints.

Unsupportable shapes

There's no gravity on the design world, but there is when you go to print. Watch for long slender sections, kinks, jagged edges, bad joins and extreme angles. Very short sections can also cause problems too if they get near the margins of what a 3D printer can express and very long ones may restrict the range of printers that will be able to create the design. If your design is going to be big then it may be worth splitting it into sections that can be physically bonded or joined after printing instead.

Aim to make a smooth design as possible and consider the orientation of print as it will be generated to avoid these kinds of problems. Your spindly starburst may look great on the screen but even SLS printers can suffer from collapsed or fragile prints that fail when they come to take them out of the powder.

The Design Software you need

What is it that you need to get going to design 3D objects? Do you need massive banks of servers in racks to the ceiling with guys rushing round in white coats and worried frowns? Not in the least.

The designing process requires the ability to build a model of the object you wish to create. This is usually done with computer aided design (CAD) software or scanning software that will generate a model. This model is then broken down into triangles often using the stl file format. This is then interpreted into as set of movement commands. The computer aided manufacturing (CAM) software[73] is used to slice the model and generate the control code (usually known as Gcode[74]) of the relevant printer This is your design 'chain', the process by which you will want to move from the idea to the printed reality.

The following is indicative of what you will need and what you are likely to find useful. Check with www.3dprintingunleashednow.com for future updates.

CAD software

Computer Aided Design (CAD) software is usually the first port of call when you want to to make a model of your idea for something you want to print. Its well worth experimenting with the different software available and get to know what you like and which are going to to be the key part of your software toolkit. Due to the varied history of CAD software there is a plethora of different file formats, some proprietary and others more open, so when thinking of your design 'chain' bear in mind that you don't want to use an unpopular program of file format which may limit what you can do with what is available to use.

CAD software does the heavy lifting of the design process and often requires Desktop PC's or Cloud services over the internet to deliver good results. Some software is starting to appear on mobile and tablet platforms but these tend to have restricted or limited options than the full blown desktop counterparts, although the gap is narrowing as the mobile offerings improve.

Open Source Software[75]

Openscad[76]

A personal favourite of mine which is great to manipulate simple shapes in a mathematical way to produce objects of precise dimension. Although the interface may appear intimidating initially it is very easy to pick up and use with a readily accessible set of help pages on the web[77]. Many of the instructions are like a simple programming language and some powerful results can be obtained from just a few lines. Another advantage of this program is it lends itself well to parametric models, where the proportions of segments of the design can change scale as needed.

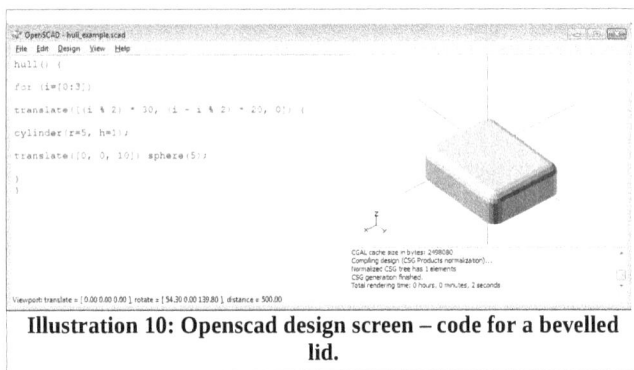

Illustration 10: Openscad design screen – code for a bevelled lid.

Blender[78]

Available on a number of different operating systems this is a scripted behind the scenes with the option to use a lot of short cuts. Blender also caters for the 3D animation market and many of its controls and options reflect this, with the result that it can be intimidating to use to start with. That said it is a powerful tool to use and well worth keeping in your 3D software toolbox.

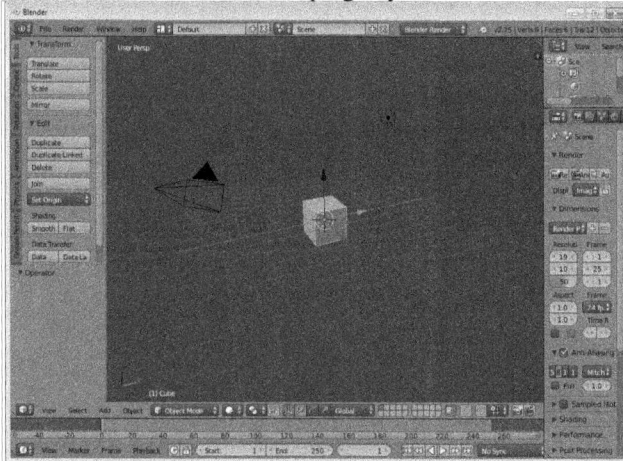

Illustration 11: The default blender screen.

Inkscape[79]

This is an open source editor for vector image graphics. Vector graphics[80] are those which use paths and nodes to describes shapes, with one main benefit in that they are often more scalable than alternatives such as bitmap (raster) graphics.

There is a cost to using vector graphics in terms of a greater workload made by the software which has to recalculate and regenerate the graphic data when changes are made. With the power of today's software and hardware this activity is often barely noticeable but used to be significant in the early days of computing with people having to buy specialist systems to cope with graphic design requirements. Inkscape is especially handy for converting flat 2D graphics to 3D embossed form, often for adding raised text onto a design face or experimenting with 3d printed lithographics.[81]

3D Printing Unleashed

Meshlabs[82]

This is another great piece of software available under a GPL (copyleft) license[83] so its free to use, with a few conditions. This is very powerful and is excellent at cleaning up the cloud point meshes generated by 3d scanning so it's able to fix errant holes in meshes and remove unwanted parts of a scan picked up by accident. It does require a certain amount of effort to master the interface and to get the best out of the software but a worthwhile investment to make especially for those of keen on 3D Scanning. A tablet version is also available too.

Proprietary Software[84]

AutoCAD[85]

AutoCAD is part of a well promoted suite of programs from the American company Autodesk, Inc who have been in the market of traditional CAD software catering for Architectural, engineering and construction since 1982.[86] They have kept up with the times and are offering their software as a cloud service too. Their software supports the DWG format[87] used by a number of different CAD suites.

Autocad 360 is also available on the android platform for tablets and phones.

Anarkik3D[88]

Winner of Several 3D awards Anarkik3D's innovative Cloud 9 design software with the emphasis on design more than engineering criteria. The software requires the use of a haptic (force feedback) mouse in order to operate and gives an impressive touchy feeling aspect to designing objects in 3D.

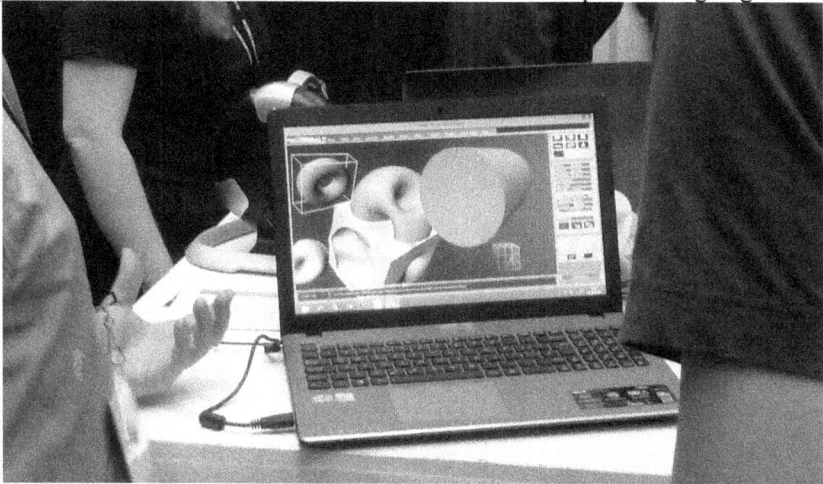

Illustration 12: Anarkik's Cloud 9 software being demonstrated at the 2014 3D Printshow

Sketchup[89]

The well known internet company that likes to cover all the angles came up with their own design software called Sketchup, originally produced by the @Last Software company in the year 2000 before it was bought by Google in 2006. However, since 2012, Trimble Navigation bought out the software in turn and has depreciated the Google part of the name. A limited free to use version is currently available called Sketchup Make.

Rhino[90]

This is a comprehensive modelling and design program which has many features that allow it to be used successfully for engineering, animation manufacturing and construction purposes as well of course for generating 3D print models. It supports many of the common file formats (including NURBS[91]) and can also be used to manipulate 3D scan mesh clouds as well. It allows for scripting to be used (Rhinoscript) and has a number of plugins available to help with particular workflows or to address specialist needs. One plugin is particularly useful in that allows you to export directly from a Rhino model to the 3D Hubs[92] printing bureau.

Tinkercad[93]

A wonderful way to learn to design for 3D printing is through the Tinkercad website. Here is an easily accessible design environment where one can get quickly engrossed in creating or adapting your own 3D model without having to download specialist software. The interface feels fresh and less intimidating for those of us just starting out in 3D.

There are many other programs and services available and the above a just a small sample worth bringing to your attention. The next type of software you will need is for slicing the model used to help generate the code needed before it can be given to a 3D printer to produce a print.

Slicing software

Slicing software is a crucial part of the process of producing 3D print. It acts as an interpretive process between converting the model's design shape and the requirements of the printer to operate to generate the desired object physically. It therefore needs not only to understand the characteristics of the model object but also the characteristics and limitations of the printer that will be used to produce the 3D print.

The software also acts to reduce the workload on the 3D printers own computer by breaking down the model of the object into fine slices and making them make sense to the printer as a set of moves and paths to follow. This involves accommodating gaps between walls, any infill, any additional internal walls required, and determining the shortest path for the printhead to follow, as applicable. The quality of the slicing is affected by the parameters the slicing software is given in respect of the particular printer you are using to produce the print. A certain level of intelligence in the software is needed in order to bridge gaps between walls and separate sections of the print and also where to generate any support for the print if this is required.

With FDM printing it is possible to retract the filament during printing. This is done to avoid drips or blobs of print oozing out and landing where they are not wanted. This may happen when the print head is moving over a gap in the print or making a long move from one area of a print to another. How this is implemented in the software may significantly affect the print quality and obviate issues with print blobbing on the external surfaces of the print being made.

Good slicing software will also take advantage of opportunities to speed up the print process, by moving the print head at maximum, speed, for example, in a section of print where no extrusion is required. The nature and type of infill selected will also influence not only how strong the final object will be but also how much feedstock or filament is required to generate it and implicitly how heavy it will end up being as well.

The software used to control these printers also underwent an evolutionary process. Skeinforge was the first generally recognised version of the software to be developed, other than an initial Java prototype, and was used to translate the 3D models into the G code necessary to run a printer. Later programs include slic3r, pronterface, Cura, and Repetier Host. These process the STL files produced by many common CAD software programs and generate the necessary G code which is then passed to the printer.

Open Source
Slic3r[94]

This is one of my personal favourite Gcode generators which has gone through quite an evolutionary set of changes in a relatively short period of time. It supports many different configurations of printer t ype and configurations with multiple extruders. New features are being added regularly, such as microlayering which allows the use of a thicker infill to speed up printing. It runs on most of the common desktop operating systems too.

Kisslicer[95]

This is quirkily named as a concatenation of the phrase 'Keep it Simple Slicer' and is available under a free license usable fro commercial use, with the caveat that it only supports a single head, with a paid for version that allows multihead printing and related features.

Cura[96]

Associated with Ultimaker, this is a suite of software to model your object, slicer it and control the printer all in one package, although it's also possible to use it as separate modules or combine it with other software as well. You can use it with most Reprap style printers or derivatives and it is reported to work well with printers using the Delta architectures (three arms). Cura also works with Ultimakers own printers with their Bowden extruder[97]. Cura now features it's own Slicing engine; previously using Skeinforge.

Skeinforge[98]

This is a Gcode generator that is based on the open source python programming language and supports over half a dozen file formats directly. It was written by Enrique Perez. It is available as a plug-in module in both the ReplicatorG and Repetier CAM software. It has proved a reliable and consistent workhorse and is associated with the RepRap project. Developments with other Slicing software appear to superseding it but it still a worthy slicing engine that produces good results.

Proprietary
Netfabb[99]

Netfabb software may come as part of a bundle with your 3D printer; alternatively a basic (demo) version is readily available with a full blown professional version for the committed.

This has some very nifty features to help repair damaged or flawed design files in a number of ways and in this respect the basic version is worth keeping handy for this purpose alone. In its full versions it be used to control your printer as well.

Autodesk 3D Print[100]

This is a neat utility to work with Autodesk's Meshmixer[101] product and allows some limited control over the slicing process as well as the standard orientation options of print placement. There are some useful features to automatically repair certain design errors and an auto thickening option.[102]

Key software to control your printer

What you use to control your printer whilst it prints can make quite a difference to your experience in using the printer. Almost all the CAM software[103] available nowadays will allow you to manage your printer, reload filament, move or adjust the print bed, and monitor the print as it progresses. You need to find a software that suits your tastes is something you can rely on so the emphasis may be more on functionality than a fancy graphic design or visuals. There will often be a recommended solution preferred by the manufacturer of your particular 3D printer. This does not mean you cannot choose an alternative to use alongside or as a replacement for that software if you feel it suits your purposes better.

RepligatorG[104]

One of the original golden oldie packages that would take your STL files and allow you to visualise your model fit in a virtual representation of your print bed area and generate the G code needed to produce a print. It features a control panel to move the printhead around and lower and raise the print bed, preheat the nozzles and reload filament as required. In addition you can update the firmware of several models 3D printer.

These are mostly Makerbot based and some other types supported, and even allowed the creation of merged G code files to produce dual headed prints, where both printheads can be used to print two different filaments at the same time. Support for this software appears to have ceased although there is a sailfish version that still continues to be supported. ReplicatorG supported the older Skeinforge slicer and in some versions of Slic3r as well. It comes in two variants, the main original tree for the software and an offshoot supporting the brilliant Sailfish firmware derivative (thank you Jetty)[105].

Pronterface[106]

This offers a reliable GUI (Graphic User Interface) with which to control your printer and allows the standard options to control your printer and manually adjust the print bed, move the printhead in various axes as we preheat the head as necessary. Communication over USB or Ethernet is supported as well as to SD cards. A benchmark for basic printer control and worth keeping alongside any other software you may have for your printer as a standby if for no other purpose.

Proprietary

Repetier host[107]

Another personal favourite of mine this software has some quirks with the graphic interface but supports Slic3r, Skeinforge and Cura slicers. It has a good preview tab and will also echo the printers build moves on the build screen with a print simulation which in itself can quite entertaining while you wait for a print job to complete. There is also an app (the Repetier Informer[108]) that can be used on a smartphone to monitor progress of the printer over a network. It has many good features and well worth checking out. Although it is proprietary software it is offered as a free download with an option to donate for regular users.

Simplify3D[109]

This is one of newer pieces of software to become available and supports a large number of different brands of 3D Printer. This package includes a well reputed slicer as well as the means to control your printer, and may be a better option than the software that came with your printer, or indeed be a version that was supplied with your printer[110].

Mobile 3D Software

There is a growing number apps for phones and tablets to help with 3D printing. These include Meshlab, Makerdroid, om3DCAD, Graphite, Autocad 360, Spacedraw, and Netfab that to name but a few. Most of these have limited capabilities compared to desktop equivalents and currently lend to viewing 3-D models as opposed to generating or editing them. Modern smartphones are becoming increasingly more powerful with a lot of interest and development being made in improving apps so it is reasonable to expect mobile platforms become more important in future for 3D printing purposes.

Chapter 6: Materials – Options and Choices

With 3D printing you have a wide variety of different materials you can use. For the most up to date information please check out the website at www.3Dprintingunleashednow.com. Something else that is important to note that while competition in the 3D printer market has reduced the price of printers, the same cannot be said of 3D printing feedstock or filament, where existing alternative uses of the materials have a strong effect on the market

A is for ABS

Going by the name Acrylonitrile Butadiene Styrene you will find most supplies using the acronym ABS. it is great stuff to print with and has a slightly waxy finish to it in appearance. Its quite tough and slightly flexible so its great to printing something that needs a bit of flex, like a box clip for example. If you drop it is hard to break. It consists of a combination of three different monomers blended to give it its print characteristics.

You may be familiar with the Lego toy bricks which clip together. These are made from ABS by injection moulding production techniques. ABS is also used in FDM printing. You can dual print it with other materials like PLA with which it adheres to quite easily when hot.

As ABS is UV resistant it survives well if you need a material for outdoor use. I designed a motor enclosure for a garden solar watering project and it is survived intact for several years out side and shows no sign of failing yet.

Available in many colours but looks somewhat dull so if you are printing a display item it might not be so suitable in appearance. You can use the solvent acetone to put a glossier finish and smooth the print out, but care needs to be taken as acetone needs careful handling and is flammable.

When heated it transitions from hard to gooey gradually and proves a great material to print with when you start off in 3D printing. It is more forgiving of temperature variations during printing and it is less likely to clog the nozzle because it usually retains its stiffness in the extruder better than most other materials.

To print ABS successfully you need a heated bed for your printer. This is in order that it will adhere to properly to the bed and not move during the print process.

If you overheat it blackens, loses its fluidity and start to jam in the nozzle. For this reason it is unwise to leave a printer idle at working temperature as it is likely to clog up. At extreme temperatures it will catch fire, by then so possibly will the rest of your printer! This is unlikely though so you do not need to worry too much about this. In the worst case with a runaway temperature in your print nozzle it will burn itself out very quickly. This is due to the low volume of material and the tendency to generate an air gap between unburnt and burnt material inside the nozzle. There are usually safeguards in the firmware that shut the printer down if high temperature limits are exceeded. A fire extinguisher nearby is a sensible precaution to take as there is also a slight risk of an electrical fire too.

One of the downsides is that it is not very environmentally friendly: It doesn't breakdown in the environment very quickly. Some people are put off by the smell. Sometimes it will warp during printing leaving it curled at the edges. This can progress during printing so much that the object comes off the print platform before its finished and ruining your print. It tends to stringing during printing, leaving very fine excess strands of filament in voids or between high points on the print perimeter.

ABS shrinks by about 2-3 % after it has been printed, which may need to be taken into account when scaling your designs. It is expected to do this as its original use was for the injection moulding process, and needed to shrink to allow easy release from a mould after injection and cooling, so the mould would be ready for the next cycle of production more quickly.

It is readily available in 1.75mm and 3mm diameters.

PLA

Polylactic acid (PLA) is made by chemically by polymerizing monomers, for example lactic acid. This process is like knitting molecules together into chains, an analogy that's sure to make any chemists out there wince.

In appearance it gives a hard glossy finish and is available in a wide variety of colours including transparent shades and often prints with vivid effect. Its great for print vases and other items where the visual appearance is a key aspect for the desired print. It is also very stiff and hard so its gives a very strong printed object.

Printing with PLA is more difficult than with ABS, due to to the way it transitions from solid to liquid. This means it stays solid and then suddenly changes to liquid, so if too much heat is present in the upper section of the nozzle then the extrusion can jam easily, also it has a tendency to drip or leak from the nozzle when printing which requires careful setting of retraction parameters to draw back the filament during gaps in the print so this is less likely to happen.

The melting temperature of PLA is much lower temperature than ABS and will often require extra fans or other cooling to help it set more quickly once it leaves the nozzle.

It is ecologically friendly in that it is often made from sugar cane, corn starch or tapioca plants, and degrades if left exposed in the environment. Some people much prefer the smell it gives off during printing compared to ABS. It is also recognised as being recyclable and is included in the category of resin identification code 7.

PLA does not warp like ABS and does not require a heated bed.

The downside is that it is quite brittle. I have dropped a fresh print on the floor and watched it shatter like glass. It also has the tendency to swell in storage if the humidity is left high, so it's worthwhile storing in an airtight box with a desiccant of some kind. It is a bad material to use to print a cup, because hot liquids will also tend to melt the cup. PLA is not good for use outside as it will start to degrade in the sunlight, unless you plan to have something you are happy to reprint every season. Available in 1.75mm and 3mm diameters.

Woody filaments

Woody filaments incorporate or simulate wooden materials. Laywood is a form of printable wood in a resin. About 40% is made from fine particles of recycled wood. It looks quite light and bright when first printed but then darkens afterwards and shrinks subsequently. It even smells like wood during printing. You can experiment with the temperature settings to get different shades of colour in the final print, to generate banding effects for example.

Because there are wood particles in the filament its better to avoid using with very fine nozzles as there is risk of clogging. It is quite flexible and leaves a rough surface.

It flows well and does not need a heated bed to print. It does not warp or shrink after printing.

One the downside its a bit expensive to buy and of limited availability, but this is sure to reduce as its popularity grows. You do need to check the nozzle size is no less than 0.35mm, 0.5mm is probably a safer bet. Available in 1.75mm and 3mm diameters. There are similar products being developed using lignin and other natural fibres, for example a product called Arbofill.

Nylon

Nylon is well known for use in many products, from stockings to garden hose attachments and ropes. It is derived from petroleum. When its produced its stretchy with a tangled up structure to the cyclical molecules and may need drawing out to get its fibres straight into a thread form with the molecules more in parallel. Known to be hard wearing it was developed in the 1930's as a silk substitute and used to make parachutes in World War II.

It's advantages are that its characteristics are well known and it is readily available and cheap to obtain, often at half the price of other materials. Some have even tried FDM printing with nylon strimmer line as a source of filament, though it should be noted that this is not an ideal source, the diameter may vary and the quality may not be entirely suitable, along with the risk of containments.

It is not only very strong it offers both flexible and wear resistant prints. It is also self lubricated to a degree and makes an excellent choice for printing gears.

Nylon is also used in Selective Laser Sintering (SLS) printing and gives a strong print with high definition finish. It is generally only available in white.

Dyeing is also possible with nylon prints, either after the print has been made or before printing by soaking the filament for a short time in clothing or tie dye. Indeed multicolour filament can be made by exposing part of a filament of a coil in one dye, then another section in a different colour to great effect. Just make sure the dye is kept hot; you can use a kettle of hot water, a bowl and some ziploc bags containing the dye and filament coil. The dying does not appear to affect the filament properties or strength of the final print.

To print it successful you either need to print on a cardboard print platform or a resin based laminate cloth. If you use cardboard you will need to strip it off mechanically after the print is finished.

It's not all sweetness and light with Nylon, however. You need to make sure it's dried before printing and kept stored in much the same way as PLA with a desiccant. If you don't do this you get a print that looks foamy rather than a clean and clear print. It is more 'stringy' than ABS or PLA and can easily result in a messy print. It has a tendency to warp even more than ABS.

Nylon has a slow decay rate in the environment and should be recycled whenever possible.

Concrete

You may have heard of the concept of 3D printed houses. This has actually been done albeit on a small scale. Many architectural firms have become interested in the potential for 3D printing houses and other buildings.

The technology offers several advantages over current techniques. The source material can be mixed on site as with conventional concrete. Where it wins is that it does not need a mould to produce a result, saving time and money. It has the potential to allow architects to design in features such as foundations for staircases, fireplaces, columns, ducting or channels for internal wiring and service conduits, and even some types of internal furniture such as kitchen islands or wall separators. These features would otherwise need to be installed separately. There is also the opportunity to have custom designs using organic shapes that would otherwise be hard to achieve conventionally or more expensively.

Some new 3D printers are coming out, either deploying a gantry type system to move a concrete extruding nozzle over a print bed, or alternately using a movable robotic style printer which extrudes along a path as it moves along. Both types of design are likely to find useful application in either printing whole buildings or components and parts of a building on site or near on-site. For 3D printing of an entire skyscraper you are more likely to want a swarm of robotic printing units than a huge frame towering above the skyscraper. An alternative scenario might be a printer that rests on layers of the building already printed, moving up with the building as it progresses.

If you do want to experiment with this kind of printing bear in mind that you need to have a bind for the cement and something like cellulose or rubber compounds mixed in to allow the mix to flow properly and aid in water retention: You would not want the cement mix to set in the extruder before you've even started to print.

The New Hybrids

Now that momentum is building up 3D printing marketplace for filament supplies, new products are being brought out to cater for this new demand. One of the results of this is that filament material more suited to FDM extrusion can now be found. Colorfabb's PLA/PHA is a good example, where the blend of two bio degradable resins has resulted in a filament that's tougher and more flexible than PLA on its own.

As one person who has used this product remarked:

"Overall I like it, it's more flexible than regular PLA and has a solid colour instead of the slight translucency. But there are definitely colours that work better than others. The Whites, Black, Sky Blue, Intense Green and Traffic Red are ones I have had good experiences with so far, while the Ultramarine Blue can not be refined after the print at all, it generates white marks."[111]

Another new filament coming soon is from the Eastman company Amphora 3D polymer which is becoming available soon. The claims to less odorous during printing and have improved results in bridging (where the filament needs to be printed across the gap between two or more supports), better print strength and a gloss finish.

Powders and Resins

The powders needed for Selective Laser Sintering (SLS) come in a wide variety of colours and viscosity. They also vary in chemical composition and in various proportions of their constituents to achieve optimal results. Not only that but the size of the particles within the mixture can have important effects too. Any particles must be distributed evenly to avoid unexpected behaviour in the final print.

The characteristics of a good resin comprise low flammability, resistance to swelling through absorption of water, and ingredients that are not harmful to the environment or us for that matter. Problems can arise when attempting to print with a mixture with very small particles is that the smaller particles may melt at a lower temperature than expected leading to sticking or fouling in the printer and built objects may suffered from defects, holes and structural weakness. The powdered composition often lends a sanded or rough finish to a print but the resolution that can be obtained can be very good and printed parts rarely suffer from delamination sometime produced by poorly tuned FDM printers.

The resins that are curable on exposure to ultraviolet light (UV) are also used in certain types of SLA printers. Here the crossbeam of two lasers generates the print point and makes the resin where they cross start to set. One beam on its own having insufficient energy to trigger the transition in the resin, so its only where the two beams converge that this happens. Colours range from clear, white, through shades of grey then to black and the resulting prints can look like frosted glass but without the same weight or heft in the hand.

Exotic Feedstock

More exotic feedstock can be used, often needing special nozzles and extruder combinations, or totally different means of obtaining prints using pumps or syringe type mechanisms. Some of these feedstocks include printing with ceramics which can be dried and then fired in a kiln. My favourite is printing in chocolate with batches of chocolate paste extruded from large syringe mechanisms. These are great for printing patterns on cakes provided they are reasonably flat or iced to make them so. Other exotics materials include sand and metal clay but these are far from mainstream processes and lend to industrial facilities and processes more then households.

Environmental Impact - recycling and reuse

The environmental credentials of 3D printing are quite good, even though they do consume energy and use up precious materials in order to operate. How can I say this you may ask? Because 3D printers can produce an object with out producing much in the way of waste materials compared to more traditional approaches. To produce one object you may even only need the 3D printer and no other machinery to refine or finish the product, so no need for lathes or drill that you might need in more conventional production. You don't need to manufacture, store or dispose of lots of different molds, as the variation in design is all in the software. Moreover, unless you are printing with ABS with a heated bed, you will find your printer only draw a few tens of watts of electricity, comparable with the power requirement of a good laptop. People have even used solar power for their prints, even in the case of one prototype fusing sand with focused sunlight in the desert to make a bowl with the focussed power of the sun as well as solar panels to power the electronics.

Another interesting potential of 3D printing in plastic is the ability to use recycled plastics. These cannot be used directly and still need to be preprocessed first to be carefully extruded as filament from pellets or ground up failed prints. PLA and ABS can be recycled in this way and there is growing interest in widening the range of plastics that get be used in this way.

Filabot[112]

From an initial campaign[113] on crowd funding site Kickstarter Filabot seems to gone from strength to strength. They came up with the idea of producing a desktop extruder to create printer filament from plastic pellets, allowing people to produce cheaper filament feedstock of consistent diameter according to their own demands and needs. The possibility of recycling used plastic filament came about later by adding a desktop shredder to produce failed prints to small enough size to work in the filament and be recreated as fresh filament. From an environmental perspective this is very laudable although filament recycling is unlikely to become a mainstream activity in the near future.

Filament Suppliers

Your printer supplier or manufacturer is likely to be more than happy to supply you with the right feedstock to meet your printing needs. Indeed with some 3D printers requiring filament cartridges[114] you may find you have on you need to stick with them. However there are many other suppliers in the market place you get to meet your feedstock requirements that will work just as well as the manufacturers brands. The following may give you a flavour of your range of options.

Faberdashery[115]

The boutique end of filament suppliers with a wide variety of different colours available and a consistent filament quality in terms of materials used and in dimensional stability.

Colorfabb

This European supplier deserves special mention because they not only provide a good range of good quality ABS and PLA feedstocks but have also introduced into the market more exotic options. They supply PLA/ PHA hybrid, their XT-Copolyester and range of plastic pellets as well.

Ebay & Amazon Marketplace Suppliers

There are many suppliers to be found especially on ebay, however a certain amount of care is needed. Some of prices quoted may seem too good to be true in some cases they are too good to be true. Filament suppliers may not have good quality controls in place and your ability to print may be dramatically affected by this. There have been cases of people finding tiny ball bearings in the filament, left over from the production processes. These will rapidly block an extruder's nozzle. It is quite common to have variations in the diameter of the filament which may cause your extruder to jam and fail to extrude correctly.

Another issue with the variation in the diameter of your filament is you may get insufficient filling and gaps in your print as well as layers that have not stuck together properly. Storage some print materials is important as they can absorb moisture which makes them swell and in extreme situations jam the extruder as well but more commonly cause bubbling to show as the moisture is evaporated as gas after leaving the extruder. It's commonly a good idea to buy a small amount of feedstock and try it before use and any feedback on suppliers and user groups or other marketplace feedback is worth checking. Black filament is particularly vulnerable as the colour can hide recycled or mixed content that is not as good as that used for natural coloured filament.

If you are unfortunate enough to own a printer that has proprietary filament cartridges be aware that there is often a way round to using the generic refills. Doing this may affect your warranty and support from the manufacturer.

Chapter 7:The Internet and 3D Printing

Copyleft and Copyright

Someone had put a nifty design up on an online repository, and set for use under the creative commons license. On visiting the 3D Printshow he spotted his print on display at a 3D manufacturers stand but was surprised to see there was no copyright sign attributing himself as the creator of the design. When he spoke to one of the people at the stand they did not acknowledge its origin. So a company that was itself promoting 3D products was infringing the copyright rules available to designers in the industry and giving bad example to all the visitors who might come away with the idea that no copyright applies to 3D Printing.

Just because many 3D designs are readily available online does not mean you can download them and use they any way you like. Moreover scanning an object does not confer any rights, as an action in itself, to the person doing the scanning, even if they have taken time and effort in doing so. It is similar to the way the law on photocopying something works. If you do or make something that is creative and that is possible to protect, then you have a good claim for copyright (enforcing it is whole other matter). Effort that one may put in terms of functionality or purpose does not count. If you scan something (with permission) and then do something recognisably creative with it then you should get some rights for the new derivative creative elements. There is no protection for something that is a copy of an original work, even if it is in a new format or was very demanding to produce. If an object is protected or patented and you scan it you are not violating the protection, but if you share the file or use it to print the object you will be infringing.

So why, we may ask, are there so many 3D designs readily available? This is because creators are often a generous lot, and are willing to share what they create with others, provided certain restrictions apply. If we look at the creative commons[116] licenses which are in popular use by millions of creators, we find the following restrictions can be applied.

Creative Commons – Attribution.

Here the only restriction is that you must credit the original creator for their work. This is the most unrestricted version of the license. It's often polite to upload a photo of your copy of the printed object or tag the repository page to acknowledge that you made one. The creator then knows how popular his design might be, who has altered the design to make a derivative and whether there is interest in developing the design further. All creative Commons licenses include the Attribution requirement.

Creative Commons – Sharealike.

This license adds a requirement that all new work must also be covered by the same license, thus perpetuating the attribution requirement. This has also been called 'Copyleft' licensing.

Creative Commons – NoDerivs.

For this license reproduction is also unrestricted, except that derivatives or modifications of the design are not allowed, so the design is effectively frozen except for changes or updates made by the original creator.

Creative Commons – Non-Commercial.

No commercial use of the design is permitted, so you cannot sell or use the object for marketing purposes without obtaining the creators permission.

These can be applied on a mix and match basis currently in six combinations. The most restrictive is the Attribution, Non-Commercial, NoDerivs combination.

Without such licenses many designs would be created on an all rights reserved and permission would need to be sought before they could be used by others. This would greatly impede their dissemination over the internet and the global use the creations could be put to. Laws in different countries vary and may override such licensing so you need to know which laws may apply for your circumstances and purposes as well.

What ideas have you got that you would like to share with others?

How the Internet and 3D Printing work hand in hand

It is not essential to have or use the internet to operate a 3D printer. What it does do is offer a way to communicate with your printer manufacturer (unless you built one yourself) to help fix problems, obtain updates of firmware and software. The option is there to collaborate on any of the manufactures support forums or communities to gain knowledge and share experience in using the printer.

Without the internet you will be much less likely to move your designs on or collaborate with other like minded individuals across the globe to a common purpose. The sheer amount of information, hints and tips available is mindboggling. It is unlikely that any activity you can think of with 3D printing that someone somewhere else has not already thought about. Without the ability to cross national boundaries and post a helpful comment or ask for help its likely your progress would be greatly retarded without this superb resource. Don't be the hermit and hide away from the world, embrace the opportunities it brings and welcome the leverage it gives your thoughts to enable you to progress.

Moreover, the internet is not just representative of peoples knowledge, it also stores many design forms for you to adapt and use for yourself. This brings us to the Internet of Things.

Internet of Things

You may already have come across the phrase Internet of Things without fully realising its implications. For those of us who are not familiar with this phrase it encompasses the concept of enhancing objects with smartness, where previously not been applied. It also leads on to ideas about smart grids and smart clouds where multiple objects inter communicate and exchange relevant information garnered through the use of the smart objects or obtained from information gathered about the environment in which the objects are located. It involves technologies related to machine to machine communication and the ability to process what in some quarters is called big data. Such smart objects can be relatively passive in their interactions, only communicating when interrogated by smarter object, or more active in having their own inbuilt processing capabilities to respond to their own sensors and information feeds. This has great potential for enhancing the way we live and use our objects and interact with the world around us.

A simple example of this is a weather station that can broadcast its data onto the net via a wireless connection and allow websites to display the live feeds or even to have services process the data gathered together to gain insight into weather patterns for a particular region or timeline.

In a hospital soon you will find that remote monitoring of patients will become the norm, with medical data recorded automatically with the potential to generate alarms and warnings if a patients condition starts to deteriorate before it might otherwise become apparent. Smart pills are already available which can report back from within our digestive system.[117]

Applications for this technology include prosaic object tracking for inventory and stock control, surveillance, healthcare monitoring[118], object and people locating and telepresence and control of remote objects. Whether these uses lend to benign purposes or not depends on how we set the technology in motion. It does not necessarily lead to a dystopian scenario as envisaged in George Orwell's book Nineteen Eighty-Four[119].

The currently available technologies are not so advanced yet that we have 'smart dust' available, so beloved of science fiction writers, where ant-like intelligence is engendered at the nano scale size of operation.

A simple way of adding intelligence to an object is to insert or apply a radio frequency identification (RFID) tag. This tag is able to communicate with your smartphone for example at short distances with radio waves through the use of the smart phones near field communication (NFC) capabilities.

One way this can work is by placing your smart phone over an RFID sticker it reads off a few bytes of data from it that tell your mobile phone to launch the browser pointing to a particular website that relates to the item. This is preset in the RFID tag so one needs to decide what to put on the RFID tag or sticker beforehand. The nice thing about these tags from a 3D printing perspective is these are small, cheap and thin and can be placed within an object partway through printing, by pausing the print and placing the sticker within the build space resuming the print which may, depending upon the design, enclose the tag sticker upon final completion of the print. This can be used to authenticate the origin of the object being printed, give an electronic equivalent of a serial number, or simply direct anyone who scans the tag afterwards to some information they might find useful such as a webpage of instructions. It is also possible to program the tag after it has been inserted although this may take longer to do if this means the tag is further away from the smartphone you are using to do this with, as the further away you are the longer it takes to communicate with.

Sharing your designs freely and building a community

We have already touched earlier in the chapter how a design can be licensed for sharing, but why do people do it? Is the motivation purely for money or can altruism play a part?

Selling your designs

Once you have a design of your own you may decide you wish to sell it to others to recoup some value for all the hard work and thought you have put in to design it. It is certainly possible to setup your own website for this or use one of the many online auctions that are available to us to sell these designs. Another option is to use some of the online repositories to promote and sell your design for you.

Repositories online or otherwise

How do you let others use your own wonderful designs? While you can share over your home network or email your friends one of the best ways of sharing your work through online repositories. Here's a few to whet your appetite.

Thingiverse[120]

One of the earlier sites for putting your designs online. Heavily associated with Makerbot (now part of Stratasys) which was originally conceived with open source principles as its main focus, it now has reduced its emphasis in that area going as far as to force users to grant the company license on the user content uploaded to the site.[121] That said there are a large number of designs available as well as community groups, contests, collections and the Makerbot Customizer (similar in capabilities to Openscad parametrised designs).

Grabcad[122]

One of the quirkier repositories available without libraries containing some designs that may not be suitable for 3D printing. There is a more CAD oriented flavour to many of the designs which may or may not be intended for printing out it remains nevertheless a useful resource for 3D designs and has a strong community following with regular competitions to boot.

Shapeways[123]

With a distinct commercial orientation to the site Shapeways has a strong online presence and is often to be found with a stall at the 3D Printshow and other exhibitions around the world. Having spoken to some of the team behind the scenes I can vouch for their friendly and constructive attitudes, which is reflected on the services they provide. They offer the option to act as a 3D print marketplace as well as repository and the ability to order prints of the designs as well. You can even hire a designer here to help further your ambitions.

Pinshape[124]

Pinshape is a wonderful sharp bright window on the world of 3D print designs. It allows you to upload your designs and sell them if you wish. Like Thingiverse it has plenty of collections in place to help find those interesting designs. As an independent repository it has much to recommend it.[125]

Online 3D print bureaus

Imaterliase[126]

A Belgian company with a strong presence at the 3D print shows who demonstrated high quality samples made from well over a dozen different types of materials, including metals. It exhibits a number of portals which allow user customised designs to be ordered so if you're after a ring made with your own initials this may be the place to look.

Shapeways[127]

Yes the US based online repository does prints too.

3D Hubs[128]

Taking community spirit that step further this site allows you to upload your design and find the nearest local 3D printer, commercial or otherwise, to produce a print for you. You can choose on the basis of locality, price, quality or size capacity. I wonderful idea that appears to be taking off like wildfire connections being made in a large number of cities across the globe. A brilliant option if you want to find out what 3-D printing is capable of.

3D Creation Lab[129]

One of the new generation of burgeoning 3D printing bureau, 3D creation lab offers a simple and straightforward way of getting your design from design file to final print without resorting to the need to purchase 3D printer of your own. They have a gallery of completed prints which give you a good idea of the range of work they can accommodate.

Fun things to print

Wind Turbines, bracelets, tablet stands, mobile covers, charms, toys, perpetual motion machines, containers, puzzles, chess pieces, Christmas decorations, shoes, Quadroptors and model aircraft, kitchen utensils, in-car holders, lampshades, robots, hooks and clips to mention just a few of the eclectic objects readily available to download. Nothing there catching your fancy? You can have a go designing it yourself or even post a request for someone to design it for you!

Chapter 8: The Future Trends in 3D Printing

What will happen next for 3D printing? It is sometimes said that science fiction can become science fact. Here is a short story to give us a flavour of the future.

Future Imperfect.

He paused, coffee cup in mid flight to his mouth. It wasn't the original favourite cup though, he had printed another copy on the flight up. There was a sudden sharp pain in his chest and he hunched over instinctively. The cloned coffee cup floated serenely away from his grasp as he let it go, its contents in time would leave an unsightly stain on the far panelling. His medical telltales in his glasses flashed red then subsided to that flashing amber again. A pop-up informed him of another heart attack and confirmed his backup heart had taken over, again.

Sighing, he checked the time bar for the restart. He wished he'd taken the extra precaution of having a spare heart grown back on earth before he left. Now it was down to the nanotech meds in his body to clear out the dead tissue and stimulate the new growth. There was nothing to be done to refresh the cells telomeres left in his existing heart though. You only get so many beats before they give up, be it in an elephant or a mouse, he thought sourly. The nanotech can help a lot by clearing out the dozy somnolent cells that just decide to sleep instead of carrying on dividing and being busy doing normal cell stuff. That leaves the active cells but still doesn't reset their mortality timers. All this thinking was of no help to him right now.

At least he'd had the cockroach inspired cluster heart installed as a backup, eight small chambers in a loop so if one of them failed he'd still have most of them working anyway. Trouble is they weren't good for long term use so his main heart had to be kicked back to work sometime soon. Maybe he should have printed out that new turbine design as a heart replacement instead, but he hadn't wanted to wait. It was also weird thinking of walking around without a heartbeat: At what point do you stop being truly human?

The retrieved what was left of his coffee and started to suck it down, ignoring the protests from his internal medics at the sudden introduction of caffeine into their molecular problems.

He swiped for some relaxing music and sank back into a more relaxed position. A com request bobbed up and he took it.

"That you Jed?", came a querulous voice from the ether.

"Yup", he said, as a man of many words. "It's me."

"You came over a little peaky on the remotes, you had one of your turns again?" came the reply

"Yup", he said again.

"Well I told you to do something about it before. Could you be bothered, nah. Too wrapped in your hoity-toity plans to look after number one. If I hadn't nagged and nagged you for a backup you'd be a gasping like a fish right now, and as blue as whale" came the formulaic reply, weary in his mind for its many previous iterations.

"I'm all right, love, you knows that" he said. Changing the topic he asked "how are my hoity-toity plans coming along anyways?"

"They be doing just fine right now. Only a few complaints from the overseer" She responded. "I told it what for, for now"

"What exactly" he said, "I thought the materials were all set up now"

"Well they were, but apparently some of them blew away in a storm and some other stuff wasn't good enough" She paused, "I'm sending you the list now"

"Okay well you take care now and I'll check the timelines from the list, you take care now"

"Bye now and don't forget the pump upgrade again, it's getting critical"

"Uh huh, bye now"

He shook himself mentally like a dog, and settled into the report as it slowly poured it way over to him following the transmission.

It had all seemed so straightforward six months ago…

"Hmph, look at this" he said and flicked the feed over so she could check it out. "What do you think?" He added.

"Well I don't know" she replied "whatever makes you think we should retire now anyway, the businesses still need running."

"We can automate can't we?" he said plaintively, "And what about young Jimmy"

She gave him a disapproving glance, "Young Jimmy doesn't know yet which side of his bread is buttered, he's only 43 to boot. Barely a spratling if you ask me."

"Well I was working out okay in my forties" he responded gamely

"Yes" she snapped, "that was because you had me to hold your hand!"

He paused and ask tentatively "So what do you think?"

"Of what?" she came back

"Living on the Moon of course" he said slowly "A bit pricey of course but sounds like it's worth it"

"Well, it depends" she replied, and added "I've set up a searcher to check out the angles" She gave him a sly grin "and of course Angela was talking about her cousin Ruth was already signed up! Can't get left behind…"She trailed off

He sat up in shock, check the logs and found she actually set of her first search queries three weeks ago!

"What took you so long?" Sheila said softly, and came over and gave him a hug.

That was just the start. Three months later they were booked and ready to go. He didn't much care for the journey but at least it was a soft and gentle rise out of the atmosphere and no one had to endure the violent rocket rides the old pioneers of space had to suffer.

He remembered on landing the warm reception they were given as they were led to the nearest cafeteria their first meal on the moon. It was strange you couldn't tell the difference between the more moon grown food and that of the earth, but as these days almost everything was that grown anyway he guessed the process was much the same up here is down there.

The rep came over to talk them over and explain there had been a delay in building their home, due to some breakdowns but that the robotics were being reprinted with a few extras and their home space would be built in the next twelve hours or so, then take a day or so to set and be tested and that if they wanted any changes from the standard layout to let them know sooner or later as modifications weren't as easy up here as down there. They nodded their heads and meanwhile checked out their own versions of the technical reports they'd extracted a few hours earlier to make notes of the last minute changes.

"Says here you've reduced the dome height by thirty three centimetres, why is that?"

"How do you know that?"

"We're programmers of course, always was always will be, what did you expect?"

"Really?" the rep hesitated, tilted her head as she listened to another conversation, and then said "Would you two mind coming with me for a moment?"

They looked at each other then both nodded. All three were soon bouncing out of the room.

The rep took them to another area full of boxes strewn around in a haphazard fashion where a young eighty year stood at a display zone shaking his fist. "Whoever wrote this I'd wring their neck", he said vehemently. "I'll give them test mode! Half of this stuff is missing! God darn youngsters have no sense of responsibility!"

"What" Sheila piped up, "seems to be the problem?"

"Hmm?" The man said, almost as if dazed, "The timing code is all up the spout, half these bots are doing stuff fine but in the wrong order, and some of the printing bots keep coming up with mud pies writing 'TESTING' on the base layers instead of a proper print, I'm at my wits end!" He exclaimed. "I'm George by the way, in case you hadn't guessed already", he added, tugging distractedly at his hair.

"So why aren't you getting help from 'down below'?" she inquired.

"Well I am but it's not helping!" replied the frustrated man. "Every time I reload their updates it still goes wrong, just the same. They're saying I'm just not doing it right."

"Did you test the checksums for the uploads?" interjected Jed

"Well of course", he snapped back.

"Well lets just see, shall we. What have they told you to do and how are you getting their updates?"

"Oh, um, its like this..." He showed them the console and what had been going on.

They were back at the refectory indulging in a bout of coffee round a small table shaped somewhat like a toadstool with a spigot in the middle from which they could get their refills.

"So..." said Jed contemplatively, "you 've been doing exactly as directed yet the behaviour of the bots shows something else at work, and downbelow are also tearing their hair out because the bots are behaving as if the updates don't work properly."

"Tell me something I don't know already" said George, looking at them through red eyes.

"Well, the code is picked up through the Earth link, passed to your main system and that all checks out okay", Sheila said ruminatively "So how does it get distrusted amongst the bots, after all some of them are all over the place and may not get a signal from here."

"Oh that's simple, we download to the motherbots who then sends out to the nearest bots to distribute the update. Sometimes it seems to work too.."

"Aha," said Jed, "is there any pattern to that, are only certain bots being affected?"

"I'll run up a quick query and run Perturbation over it" said Sheila

"What's Perturbation?" asked George.

"Just one of Sheila's pet neural nets she wrote up a while back, should be quite quick" replied Jed

"Ah look at this" said Sheila, "there's a couple of bots always seem to be involved, 41687 and 99832, see?"

"Oh WOW" George exploded, "that's like incredible, what's going on?"

"Why don't you recall them and check them out" said Jed

"I can't, that will kill the production" George blurted. "They're on the main nexus"

"Well it looks like production isn't doing too well as it is, is it?" said Jed gently.

"I suppose not" said George dejectedly, "but if it's not fixed soon it'll hit lifesupport"

"Can't we substitute some of the others or print some spares, you do have spares don't you."

George looked at them with a pained expression, "I guess so, but its not going to be easy to modify what spare and I don't have enough resources yet to reprint them, we're on a knife edge right now"

"Well let's haul them back in and see if we can do a repair"

It took a while to get the two bots to come back for inspection, as they seemed resistant to the recall order and it took several goes. Finally they crowded round in the small round maintenance bay that appeared to have just been finished a few minutes ago, George claimed it was fine as it was over two hours old since it had been built and there weren't any leaks.

A faint actinic smell hit them as they opened the air lock and the diminutive dusty shape of 41687 shuffled forward on its partly inflated soft legs. 99832 was waiting outside still as despite their small size there wasn't room for two of the bots in at the same time, unless they deflated fully in which case they would no longer be able move by themselves.

George went up to it and ran a standard diagnostic on it from his mech Pad. He looked up puzzled.

"That's strange, it says nothings wrong, look it's passed all the tests, see" he said and showed the other two the display on the quaint looking device.

"Hmm", said Jed "Can I have a hold of that pad a second"

George handed it over gingerly, "be careful with that, it's the last one left"

"Why is that?" Sheila asked carefully.

"Oh, we had a solar flare a while back" George explained casually "the others were exposed and got fried."

"What makes you think this one's any good either?" Sheila asked.

"Well it's working now, isn't it?" said George sharply.

"No I don't think it is" Said Jed, "look I've just run a small memory test on it remotely and it doesn't tally, see"

"You're kidding!" said George "here let me see"

"I reckon you've a bad batch of memory knocking around, what do you reckon, Sheila?"

"Hmm, looks that way" she replied. "We'll have to rig up some way of testing the bots in the field and blocking out the damaged areas, unless you've got plenty of spares George?"

George just shook his head sadly.

"Let's get to work then" said Jed with a certain glee in his voice. "Who knew retirement could be such fun, eh" he added to himself.

It took several hours to make preliminary code and deploy to the remaining bots to see how extensive the damage was. It took several days to get things organised so that the semi lobotomised bots could function as a team, with revised code from 'downbelow' that forced memory checks every so often, while they also worked on a redesign of the bots to make them more resilient. It was several weeks before they could ship out extra parts due to an insurgency at the spaceport that took a while to die down, so they had to manage with what they had to hand. Some of their compatriots donated spare tech to help with the cause and soon the bots were working something like normal. It meant some had to be on hand to monitor what was going on and be in place to judiciously swap in and out various code segments as many of the bots no longer had the capacity to hold all the code they needed to use in one go.

Six months later and they had been doing quite well and were just getting close to break even on many of their supplies. The hydroponics areas were now built, the 'green sludge silos' as Jed liked to call them.

The moon retirement colony was going to be a success. They were printing bots, they were printing domes, and they were printing food blocks. Bots were scouring the moon for water deposits and key minerals as well as gathering up regular moon dust to be fused and printed into domes and internal fittings.

Jed was helping out with the field monitoring in a small air pod they had built near the bots operating area. It included a coffee machine, printed out especially at his insistence, along with a copy of his favourite mug.

He quite liked sitting out there and juggling the needs of his bot herd, as he started to call them. He had just been sending a report back when he had a glitch with his heart in mid slurp. It had been touch and go and they had rushed out a Bot to bring him back to base safely. It made a mess of the air pod though when tugging it back with him still bumping around inside.

They ended up printing Jed a new heart, grown from his own cells, but it all took a while. At least he got to keep a hear that beat and not a turbine. What happened next though, was a whole other story.

-o0o-

I think what we can take from this is that 3D printing technology will become so mundane and integrated in our lives we won't think of it as anything special. What directions is it going in the near future? Lets look at some trends.

The trends guiding the future of 3D printing

3D printing is a tool that is constantly evolving and changing. More and more applications are opening up as people realise the potential for using it in new weighs, from printing with plant seeds to making a precise model of a baby's heart to help in an operation on the real thing. 3D printing's origins were as an aid to prototyping, now more and more of 3DPrinting output is being used as the final product.

One of the trends that 3D printing enables is the growth of customised products that is products tailored for a particular person or specific need. 3D printing drastically reduces the tooling costs that might otherwise occur with more conventional production processes.

The speed with which a printer operates is also likely to improve in the future. Currently you have to wait minutes or hours for a print to complete, often from the output of a single nozzle. Improvements in extrusion delivery, the use of multiple extruder's akin to the way Ink Jet printers use multiple nozzles at the same time will all mean a reduction in wait time to print. One can't defeat the physical limitations and requirements to actually move a printhead in time or space, but we can make the printing process more efficient by say having nozzles that can dilate to different output sizes during the print process, without having to change over the printhead using a turret to swap the nozzles round. Such a printhead would be able to restrict extrusion to a fine diameter in printing the external detailing and maybe expand the diameter when printing the infill, the area inside the print shape, much more quickly. The coarse quality of the infill material would not be apparent when examining the external surfaces of the print, while the strength and density of the print would be comparable with one printed more slowly with a fine nozzle throughout.

The preoccupation with having to slice a design before printing it out may also fade, and allow smooth continuous printing to be achieved without the granularity of the ridges produced by the slicing effect particularly evident in the FDM type of printing. It is likely that techniques will develop that allow the printing of an object at many focal points simultaneously to achieve this effect so that the printing process becomes more of a 3D process than composites of 2D slices as is currently the case with many of the processes currently available.

Another factor is the rapidity of the production items that are produced either partly or entirely with 3D printing processes. There is no need for a new production line be built of factory setup to start using 3-D printing installation for a new product. It may be necessary to alter the mix of feedstocks and add posts printing production processes into the mix this can be a lot less time-consuming than the more conventional alternatives.

The option to mix and match means of production is also becoming apparent, whereby the different production methods can be used during a products lifetime, so a product may rely on 3d printing initially and when production scales use injection molding of some of all the production runs. Alternatively hybrid production becomes possible, where some finishing or machining of a 3d printed part is applied in order to obtain the desired result, using the best of both worlds[130].

Scale will also play a factor, 3d printers will increasing be seen operating on the large scale, for infrastructure and architectural projects, to the very small nano sized production of say graphene based circuitry and nano scale robotics. The range of what can be produced will expand as well as the range and type of feedstocks we will be able to chose from will increase.

Cloud robotics will also play a part in the development of swarm printing. Have you ever seen a picture of termite mound and realised that it has all been built by huge number of tiny insects? The same multiplier effect will be possible with swarms of robotic mobile 3D printers, able to co-operate to build objects much bigger than themselves and much more rapidly than a single dedicated machine. If one of the swarm fails or is broken it can report back and either be replaced or repaired while the rest of the swarm carry on building. The swarm members will be able to communicate between themselves on their location and progress in producing the design and will be able to retrieve or request fresh feedstock when they are running low or report back on unexpected problems.

This flexibility of 3D printing means that novel items can be produced and tested in the marketplace with without incurring costs entering or generating new market. This gives more freedom for individuals and firms to experiment with the kind of things people might want and quickly obtain feedback as to whether or not the idea for the product is likely to be successful and allows any modification of the product to be made swiftly and less costly to align with the demands of the marketplace. So if for example producing a product in a variety of colours it's very easy to change the proportion of production from particular colour in response to changes in demand which may occur as a result of changes of fashion, season, or for other more intangible reasons.

A convergence is apparent in the fields of 3D printing and robotics. A 3D printer is after all a type of industrial robot in itself. Robotics is blossoming in its own right but not only in terms of large human -sized robots or heavy industrial robot machinery. It was also blossoming in terms of tiny machines that can work on nano scale. Ant-like behaviours of robot swarms will also feature strong strongly in the future and the ability of 3D printers to produce a production run of such and like robots in a variety of configurations is likely to feature prominently in the future.

Let us see what implications these trends have for us in the near future.

Wearables/Personal Items

When you go to shopping you will find a variety clothes shops reflecting different fashions and purposes. Moreover many of the clothes that you see on the myriad racks are of the same design but different sizes none of which precisely match anyone individual to a high degree of tolerance. They are all approximations for clothing will be a good enough fit to be suitable for use. Now consider the kind of impact 3D printing will have. You will be able to walk into a body booth, get scanned and be fitted with the latest gear that will be an exact match in colour and shape for your needs. No more one size fits all mantras.This will mean the need to stock huge varieties of the same clothing in a range of sizes will no longer be necessary. This will make life simpler and reduce costs. The customer will also benefit from greater utility derived from the clothing that they acquire.

Biological/Medical Uses

In these days of mobiles and smartphones I know of friends who say to me 'what do I need to wear a watch for, there's a clock on my phone so why do I need to carry something else?' Now thinking of the future I can almost hear people in the future saying 'what do I need to carry a phone for, there's a phone in my wrist so why do I need to carry something else?' I reckon people in the future will be heavily into augmenting themselves with implantable devices, such as a phone. Where does 3D printing fit in with this? Well most people are physically different and not just by the size of their feet. For implantable devices to work well they are likely to need to be shaped customised for each person, and I envisage that 3D printing will help considerably in that.

Illustration 13: Noa Raviv's wearable designs on display at the 2014 3D Printshow

3D Printing Unleashed

The flexibility of 3-D printing will also play a part in this by allowing people to choose the own particular mix or preference of implantable devices to meet their needs or desires and cater for one-off or rare combinations or permutations of a set of designs or functions.

There are recent medical advances in growing cell structures by generating suitable structural raft in which a batch of cells can be inserted. The structures allow the cells to organise themselves in the required order to work properly as a particular organ or body part. By taking cells from us it may be possible in the future to grow replacement or spare parts with which to repair ourselves which suffer none of the rejection issues that may occur with transplants. Such methods will also increase the supply of medical organs can be used without relying so much upon a steady stream of donors who are currently in short supply. 3D printing will be able to play a part in this print structural rafts required for particular medical needs and to match individual body configurations.

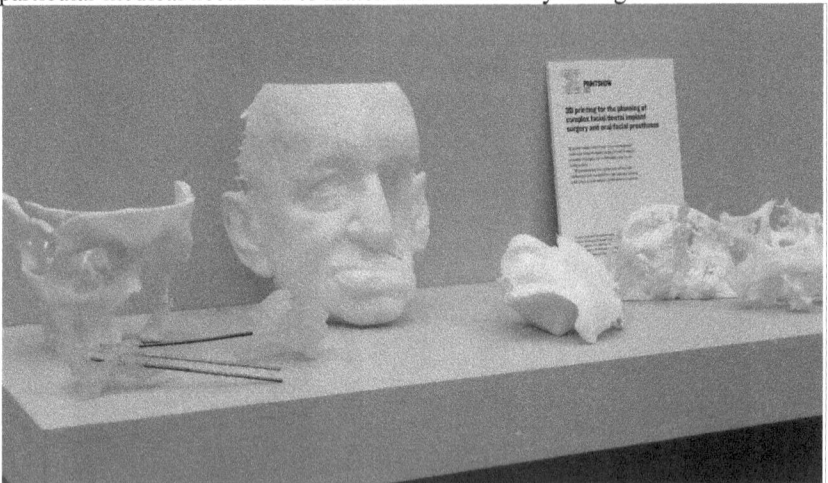

Illustration 14: 3D prints used to assist in planning medical procedures

3D printing has already been used to assist in preparation for operations through printing out the results of body scans to allow surgeons to visualise procedures they need to follow for particular operation[131]. In the future it is possible customised robotics on the nanoscale will be generated and used inside the body to achieve specific medical goals. They could be programmed to repair damaged heart from the inside, for example, rather than having to cut it open to achieve the same result conventionally.

Maybe you have seen an episode on Star Trek where one of the actors walks up to a cabinet, asks for a meal, opens a door and the tray of food materialises in front of them. Well, 3D printers can not only print the tray and cups but also print in food right now so its not too far fetched to imagine on demand food made to order and delivered in front of your eyes. The concept of printed food is not so far-fetched as we already have printers can print the food pastes already. 3D printers will produce the organic structures need to organise and support cell structures and allow them to grow into desired shape and function so the days of having a vat grown printed steak are on the horizon.

Building and Construction

Imagine an earthquake has hit your house and demolished it. You may end up in your garden sitting in a tent while you watch your house being rebuilt by a 3D printing robot to the exact same specification used when you first printed in place originally. When you re-enter your finished home precisely the same layout and fixtures that you print it originally. It may even be a feature on alteration to the original design, perhaps a stronger roof, you decided to apply to the design package before you instructed the robot to rebuild your lost home.

The use of 3D printers to produce buildings through concrete extrusion has already been demonstrated, indeed the WinSun Decoration Design Engineering Co. has created a five storey building using 3d printing techniques using recycled building waste. In the future this will become more common and you will be able to have a house or building made to your own specific requirements or to fit the limitations of a particular plot in size and shape, with savings in production time and costs.

3D in Space – Moon & Asteroid Printing

We have all heard of satellites in space and various missions to different planets in the solar system. The majority of the systems used in the payload are automated and operate with their own computer systems.

Printing space is already a reality with a 3Dprinter having been sent up to the International Space Station (ISS) platform. Mission control sent up the design for a spanner which was then printed out in the habitat of the NASA environment successfully and then went on to print another 21 different designs to test out how well it performed[132]. The very first print was of one of the printers own components, perhaps in homage to the Reprap ideal of self replication.

The impact of this development may well lead to the design and development of robotics mixed with 3D printing capabilities to allow 3-D printing on the moon and on asteroids. Why might this be desirable? Well getting material up into space in a rocket from Earth is very expensive and time-consuming and it would make great sense to be able to use the materials already in space to build objects for our purposes. This means we would need to design mining robots that will obtain the feedstock, needed by the 3D printing processes, from moons or asteroids in situ. Imagine sending a batch of robots off to the moon with instructions to mine the lunar soil, and use it to print out habitation modules on the lunar surface and build the necessary machinery and devices to sustain life on the moon! It might take a few years for such a base to be created but during that time no direct human intervention be needed, simply monitoring from Earth as the process unfolds and amending instructions as necessary.

It is not just a moon base that could be built in this way. Using the materials from asteroids to generate the fuel needed for rockets or other spacecraft is possible. All without calling upon earth's resources which need to be brought up the gravity well. Of course, should one of the robots breakdown, we should be able to print a replacement on site without waiting for another shipment of supplies and thus reduce costs and increase our flexibility in delivering space projects.

Parallels to Other Industries.

If we want to see how the 3D printing industry is likely to develop we need only compare its developments with that of this 2D printing industry, that is printing on paper for example. The parallels are there in terms of the early printers being very expensive and cumbersome to use, such that people would resort to using print bureaus to deal with the complexity of print runs through their own specialist equipment and expertise. Nowadays 2D printers, whether laser or inkjet based, are so ubiquitous that the need for such specialised print bureaus has greatly diminished. Moreover the economies of scale mean that today's 2D printers can in some cases be bought for less than the price of the ink used to print with. However, with regards the high cost of ink in the 2D printing field, feedstock 3D printing is less restricted there being existing alternative markets for the feedstock which the 3D printers manufacturers have adapted to take advantage of with their designs. It is interesting to note that the feedstock materials are being specifically tailored for 3D printing purposes now that there is sufficient momentum in the market to make it worthwhile.

Impact on the world economy

By reducing the number of constituents needed to produce an item you reduce the complexity involved in the production process. It follows that this also number of parties or companies required to cooperate in a production sequence to generate the desired product.

The storage costs transport costs and handling costs of managing the products will be a lot less. Because you will be able to produce the products on site or near to site where you are selling the product, you will shorten the supply chain.

In the future a 3D printing clothing booth will only need the right proportions of a standard feedstock mix to cover vast array of different designs and patterns. There will be less need to transport vast numbers of different goods from one side of the world to the other. A reduced set of standard feedstocks will be needed instead. These standard feedstocks will be in high demand and therefore economies of scale will apply. This is likely to reduce the cost of making products even more.

Chapter 9: Buying Your Own 3D Printer

The different option open to you.

Given the myriad of different competing manufacturers let alone design architectures out there it is useful to have a framework to build your short list for the printer you want to buy. There are some key decisions you can make which will make the process easier.

What size of prints do you want to make? Build volumes for many of the printers available vary quite a lot, from a few centimetres tall for a portable printer to something man height. Another consideration is in terms of size is what is that you want to print? If it's jewellery then you might consider the laser resin printer type, which is excellent for fine detail work. There are some FDM types that come close too, and may be a better choice for wearable items. You may need to think too about where the printer can be put, easy with a small desktop model, but a two metre delta high may need the equivalent of a cupboard or shed space. It is even possible to get a Cartesian style printer that is about the size of a shed.

Which material do you wish to work with, plastic, metal, wood or ceramic? See the chapter on materials if you want to recap on the different types available. Some types of printer can work with several different types of feedstock, others are less generic. There is one brand of FDM printer, for example, that locks you into using their own cartridges, a bit like the way 2D ink jet printer manufacturers require you to use their type of ink refill.

Are the Electronics important, you may ask. Given the electronics in a computer strongly affects its performance, this is not the case so much with 3D printers. Many of them are based on open source computers boards, such as the Arduino range which are 8-bit computers, so the speed of the CPU doesn't bear a direct relationship on how fast your printer will go. Far more important is the quality of the firmware and the features built in. Many FDM printers being sold often have heritage from the open source Reprap project and will use firmware based on Marlin, for example.

Firmware upgrades can make a radical difference to the performance, as I have seen myself with Sailfish firmware (thanks to Jetty and Dan Newman for their sterling work) which can be installed instead of Makerbot's standard issue firmware. The result is faster prints and a cleaner finish to the final printed object. The Sailfish software was developed as an open source project. They have even expanded the range of supported printers beyond that of the Makerbot range although this does not include those machines using the Reprap style of firmware (e.g. Marlin).

The physical speed of the printer will vary according to the type of printer it is and the configuration of the extruder, the stepper motors and other architectural features. Physical speed is not necessarily the sole characteristic you need to look for, you need to consider the printers resolution and build volume. How important these factors are will depend glaringly upon your intended use.

What to look for in your printer

3D printers are physical things to do physical work. As such they will be subject to considerable wear and tear. You will need to ask yourself how much do I want to be a designer and user of a 3D printer or will I become a 3D printer maintenance engineer as well. If you do buy a printer you will almost inevitably have to do some maintenance as well, even if its only reloading feedstock and cleaning up and finishing printed objects as well. That said, there are degrees of this and you can reduce the need to wear the maintenance cap through careful choice of the printer or printer kit you buy. Sadly it is not a case that the more you spend the easier it is. Even people using industry grade 3D printers find leaving a print running overnight can result in a half made mess in the morning. 3D printing technology is still at its early stages and many machines still require a lot of effort to keep them working effectively.

For most of us I suspect we will want one ready built. You don't need to get your hex keys dirty and all the fiddly business of putting things together is done for you already. That said if the thought of Ikea-like complexities does not daunt you and you feel you have 'engineers hands' then you are likely to want to satisfy your curiosity of where each bolt and panel goes, and how it all fits together so neatly. Indeed, if you do take this route then almost any maintenance you need to undertaken can be undertaken with little dread.

One thing to consider is the weight of a 3D printer. The Leapfrog Creatr printer I received one morning turned out to weigh some fifty Kilos and it took some effort to get it over the doorstep let alone up the stairs!

If you are minded to buy a kit then it is a good idea to check what sort of problems people have reported either building or using the machine. You will want to know how vibrant the community is in supporting the printer and how much help you might expect from the company you buy from and in the community groups or channels. Have a look at how people respond to a question that's posted. Bear in mind many of the FDM models may have Reprap origins in their design and firmware and may therefore benefit from the huge wealth of knowledge and even third party add ons and parts out there.

Even if you are not so handy putting things together it shouldn't take very long. A couple of days more than a couple of weeks especially with the ready availability of parts that a few years ago you had to make instead. For example there are quite a few ready built extruder assemblies to be had on auction sites that will work with many of the commonly available printers out there whereas before these had to assembled from parts.

To buy or not to buy – Do you need to buy one at all?

One thing to bear in mind when considering buying a 3D printer is whether you need to buy one at all. After all, in can be quite an investment in time and money. If you are more interested in working in 3D Design or you are not sure you want to commit heavily in 3D printing there is are alternative options. You can use bureau or even collectives of 3D printer owners to do the printing for you. The downside is often longer lead times to get your hands on your printed design, and higher costs than doing it yourself. Of course you may do both, have your own printer but for certain types of print or materials using someone else's equipment can make sense. If I need to print in sintered titanium it makes sense to use a bureau than invest in thousands of pounds or dollars of equipment for a run of one or two prints. If you need to scale up then you might need to buy equipment later to ensure security of supply and reduce costs. Don't forget to factor in postage and delivery costs in your decision making.

For an up to date list of 3D printer suppliers please check out the website at www.3Dprintingunleashednow.com, the numbers of different manufacturers is changing every week. Here is a summary below of some of the more common ones available at the time of writing:

3D printer manufacturers

Makerbot [133]– US based – owned by Stratasys as of June 2013

Makerbot started in 2009 with an open source design for a 3D Printer, the Cupcake printer made with a chassis of laser cut wood and available as a kit. They went on to produce the Thing-o-Matic which has some improvements to the nozzles and updated firmware. These designs were all open source (freely available and modifiable) and you could download the designs yourself, cut your own chassis yourself rather than buying a kit of parts. Soon the Thing-o-matic was supplied ready built. One of the innovations of the Thing-o-matic was it's Automated Build Platform (APB) which Makerbot patented but didn't implement it well enough to become commonly adopted (though I still use mine, with certain modifications)[134].

They are well known in the 3D industry and have had a strong influence in the recent history of 3D printing with the careful cultivation of Designers through their Thingiverse online repository. There were some adverse reactions in some quarters when they were taken over by Stratasys with some designers (e.g. Dizingof) withdrawing their designs from the repository in reaction to the move.

Their current range includes the desktop Replicator series of printers and a Digitizer (3D scanner). As some of their older printer designs are still available as open source some clone manufacturers (e.g. Flashforge) use these in their products or derivatives and may arguably be better value, at the potential loss of mainstream support.

Offering good support and a wide range of models they are a sensible manufacturer to pick and provide a great deal of reassurance with the backing of a strong company behind them (Stratasys) and a spare parts bin to boot.

Ultimaker[135] - European based operating since 2011.

These guys took hold of the Reprap model and shook out a new open source design. A Dutch company from 2011 that, like Makerbot, had its initial design with a chassis of laser cut wood and originally only supplied their printer as a kit. They now sell ready built printers called the Ultimaker 2 as of September 2013 which will cheerfully print both ABS and PLA filament. While it is similar to some of the Makerbot designs, with both companies focusing on Cartesian configurations.

Ultimaker makes use of a Bowden extruder, which separates the heavy stepper motor and gearing from the nozzle, using a Teflon tube to make the connection between the two parts. It works a bit like a bicycle brake cable. This means that the Ultimaker has a very light print head and nozzle. The advantage of this is that the head can move very quickly. When I have watched an Ultimaker at work in one of the 3D print shows I always came away impressed out how quiet and smooth its printing operation appeared to be. Like the early Makerbots their printer uses an Arduino computer to control the electronics.

Stratasys /Objet[136]

Founded in 1989 this company was one of the first companies to start on the 3Dprinting path, purchasing a number of 3D related IBM patents on the way. As a dominant player in the market it took the technology from it rapid prototyping origins to the 3D printing world we have today, acquiring Objet in 2012 and Makerbot in 2013. It has a strong lineup of printers and its presence is felt in many different fields of Application.

Leapfrog[137]

Typical of one of the many start-ups in 2012 this Dutch company was an offshoot from AV Flexologic who specialise in packing machinery. Given this origin their first printer was a robust metal box with industrial components based closely on a RepRap cartesian model, suffering a number of teething problems from the rush to enter the market. They now have a wide range of extrusion based printers with much improved firmware and design features and a growing base in the community.

3doodler[138]

A kickstarter[139] community funding success story for a pen based hand held 3d printer. A cheerful presence at the 3d shows usually with hordes of enthusiastic youngsters madly printing away. Now on its second generation model I'm glad I backed them.

Community funded projects

A community funded project often seem great and it is easy to get caught up in the enthusiasm for the end result. They are much more risky than a normal purchase though as the product won't be ready straight away. You are paying for the product development as well as the product, and if the development fails or gets sidetracked you can lose your money, the product may not turn out as you expected as the specification or targets may get changed. There are often long lead times, several months is common, and its easy to get impatient.

Having said that, it can be great fun knowing you have directly contributed to the creation of a new project, which without your help might not have seen the light of day. I had personal experience of this after backing the development of the 3D Doodler[140], the first hand held 3D printer.

What to look for in selecting your 3D printer

Owning a 3D printer is not always a press and play experience. They can be tricky to setup and will not always produce reliable results. With a little care and patience however, you will be amazed at how much fun you can have with them and how rewarding owning a 3D printer is.

You will need to consider where to put your printer in you home or office. They like to be kept warm and dry, away from draughts and have a good clean power supply. You may want to consider using a small uninterruptable power supply (UPS) for your printer and PC, laptop or tablet you are using to control it.

You will need to consider the following factors:-

Technology

Do you want to use Fused Deposition Modeling (FDM) or Selective Laser Sintering (SLS) perhaps?

Size

Will it fit in your office or lounge? How big a build volume do need to be able to print, car sized or model miniatures?

Weight

Is it very heavy? Will the floor support it? Is it going to rattle the china when it runs?

Accuracy

Do you need fine details for your application visible with a magnifying glass or does the odd splodged lump not bother you at this scale because you're looking to print large scale?

Origin

Do you need a local dealer or happy to it shipped across a continent?

Support

Do you want to rely on the manufactures support forum or go with the community, instead of or as well? The popularity of the model in question may make all the difference when you need help with a model specific problem.

Speed

How fast can it print or are you more concerned with a high print quality or accuracy.

Cost

With 3D printing high cost does not automatically mean high satisfaction. It pays to do some research before reaching for your wallet.

Reliability

3D printers all have their foibles, but some are better than others. Check out the support forums and user groups to see how others are faring. A heritage of previously successful models is a good indicator of a good iterative progressing printer design.

Ease of Use

Do you want to be in total control with all the options or just press start and trust the process.

Materials

FDM, powder or Resin, which matters the most? How about chocolate, food or Biological feedstocks for more exotic options. You my also want to check if the printer comes with an enclosure if you plan printing with ABS.

Kit or ready built?

How much do you want to get your hands dirty? Kits builds be rewarding

Modifications

Is there a particular purpose you have in mind and does it require specific modification for a 3D printer (goodbye warranty) which doesn't seem to be available? It's likely someone else may be thinking along similar lines so don't forget to check out the likes of community funding sites like Kickstarter[141] to see if someone is already offering what will meet your needs. If not it may be worth sounding out your ideas on community group sites before taking the plunge (or hacksaw).

Reviews

Has the printer or manufacturer had good reviews or feedback from users? Don't just rely on a brochure to be informed about your impending purchase. Visiting a 3D Print Show and seeing one in action can be captivating but is unlikely to reveal a long term reliability issue with the design.

Best wishes on choosing a 3D printer that suits your needs, and remember there's usually a good second hand market for used and older models.

Chapter 10: Tips and Tricks

Tools to help you

There are some simple tools that we can acquire that will help us considerably in using and maintaining our 3D printers. Such tools you may want to consider are things like Allen keys to adjust the various nuts and bolts are likely to be used to hold your 3D printers together. Similarly screwdrivers and spanners of various sizes will also prove useful. I found I often use extra long nosed pliers to hand pull away excess filament or make fine adjustments.

As part of maintenance for my printer we need some lubricants. Instead of fine oil or lithium paste I recommend using a silicon PTFE grease in a spray can. This can be applied even while the machine is working should a squeak come to your ears. With a straw adapter plugged in I can spray just a small highly defined area rather than coating the whole area. It is important to avoid contaminating the build platform with any lubricants otherwise you prints might not stick to it properly. There may be specific actions needed for the particular model of printer you are using, so be sure to check on the manufacturers recommendations and advice.

Keeping a roll of adhesive Kapton tape spare is a good idea too. It's an excellent insulator and can cope with temperatures well into 250° C without scorching. It's available in various widths including an extra wide roll that can be a half or a third the width of your print bed or platform. You can use it on the bed on top of the existing surface to improve the adhesion of your prints and will only need replacing after several uses.

Again for ABS printing having some acetone available is also extremely useful. Use it to clean the surface of your print bed, you can mix it with a little bit of ABS wastes to create what's called ABS juice which spread thinly on your print bed improves adhesion, and for the brave we can also use it either brushed on our finished print or in an acetone fume bath to spend the print in an acetone filled jar to smooth out the print all the way round. Be aware the acetone can buy on the high Street is often impure and can contain oils added to improve nails when used as a nail varnish remover. This oil blocks the effectiveness of the acetone for 3D printing purposes and should be avoided. You will find your bed will not stick and any ABS will not melt your ABS juice if you use impure acetone as your solvent. Also be recommended are some microfibre cloth's which are brilliant in picking up dust of the print bed all those little bits of leftover plastic seem to spring up after several print runs.

It is worthwhile checking your favourite online repositories for any 3D print designs to enhance or improve your specific type of printer. These can be holders for tools, lubricating eggs for filament, modified extruder components, an energy chain to tame some cabling, or simply something to enhance the look of your printer or personalise it in a way that appeals to you.

Lastly it is a good idea to get one or two of the small wide bladed paint scrapers that you can get to lift off because prints that have stuck too well to your print bed. Make sure they are not too sharp as you do not want to cut your print or damage yourself in using them. With a heated bed it is often easier to wait for the print surface to cool down before removing a print as the resulting shrinkage may make the printed item easy to remove.

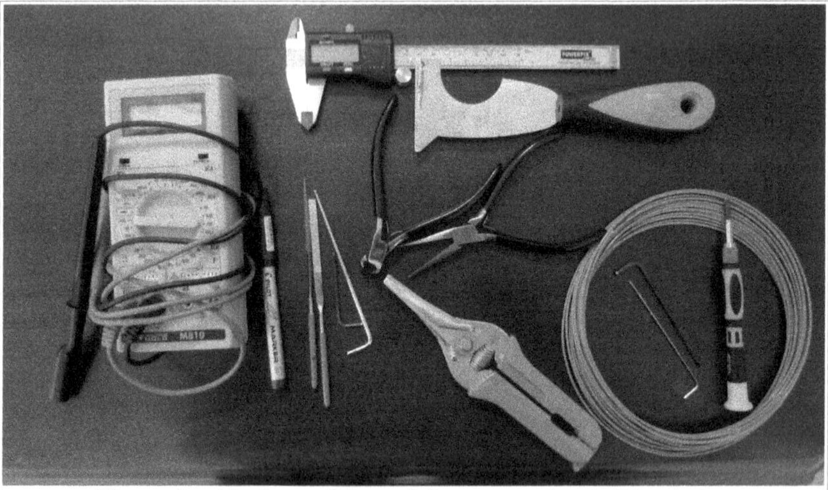

Illustration 15: Some of the tools I use most often in 3D printing

Remote controlled 3D printing

Sometimes you do not want to be hovering over your printer waiting for it to finish, but might still like to keep an eye on its progress. You can do this with a little remote control magic. Sometimes this is built into the print if it has its own Ethernet or Wifi connections and a service setup that you can log in to remotely. Other times in need some extra hardware and software.

If your printer does not come with this there is another solution by using Octoprint[142] which requires a networked Raspberry Pi computer and connects via a USB cable to your printer.

The Raspberry Pi is a favourite of mine. A cheap credit card sized device with just enough on it to make it usable as a stand-alone computer. A great feature about this machine is the excellent support that people are giving it in terms of software and hardware accessories.

One of one of the best ways to take advantage of this for 3D printing is to load the excellent Octopi server software on it, boot it up then hook up the Pi to the USB port of your 3D printer. You can then connect to it over the network as a remote printer controller which is presented by a web based interface, very similar to many to the ones you are likely to use on your own desktop.

It can now act in lieu of your PC or laptop and deliver pre-built Gcode commands to print out on your printer and control the print process over the network. This means you can have your 3D printer in one room and your laptop and PC in another with all of the advantages of being able to work remotely, plus once you've set a print off you can leave it running and switch off your desktop or laptop and go do something else.

Another neat feature is that you can also set up a web cam or the special raspberry Pi camera module itself and ask it to take regular snapshots of your print progress so you can check up on what's it's been up to. This is great for diagnosing problems or keeping records of how the print went and you could even use the images to create time-lapse videos perhaps.

If you don't like a cable runs you can always add a Wi-Fi module onto the raspberry Pi and use that instead of having a standard Ethernet cable. From an economic point of view, given that large 3D prints can take several hours or days to print, the low power consumption of a Raspberry Pi can save quite a bit of energy compared to running a desktop non-stop for the same period of time. This is such as cheap solution there is no reason why you shouldn't keep one around just in case you feel the need for using now and again. You don't have to use it all the time.

USB cables

USB cables are commonly used to control many of the 3D printers that are currently available. If you need a run of several metres between your PC and the printer itself and you may need to consider the quality of the cable you use and whether or not put additional ferrite filters on the cable to reduce the noise. You will quite often find with a long cable the number of USB errors starts to rise. This may even lead to the print stalling as the to and fro communication between computer and printer go out of sync. In the case of Repetier Host it allows you to restart the stalled print (the OK command), but otherwise you may need to abort the print altogether. It may be possible to redo the print where it left off by manually editing the gcode to do this, but this is a bit tricky and hard to get it to start again part way through a print properly.

You can avoid this by using the above Octopi solution and paying a little bit more expensive high-quality USB cables, or avoid printing over USB at all. This is done by putting your print files on SD cards and physically inserting them in the machines card slot and running your prints of the control panel, that is provided your 3D printer has these features.

Prevention and maintenance.

I strongly recommend putting a smoke alarm near your printer, especially for the FDM types. There can be a lot current going to heated beds and extruder elements and certainly enough to burn out the wiring. The print head is heated and the nozzle itself will be regularly set to oven temperatures. A fire extinguisher nearby might also be wise, provided you know how to use it. Don't leave your printer running for long periods unattended, especially in a domestic environment, even if you are monitoring it by webcam.

Lubrication is important. Bearings wear out, rods get grooved, your printer is dancing non stop during printing and can wear out the floor! I recommend silicon grease PTFE to keep things sliding and prefer it over more traditional oils or grease. Bearings also need attention now and then, but be guided by your manufacturers recommendations.

Your print bed likes to wear a coat. People have successfully printing on Kaptan tape, clear adhesive sheets, glass panels, ABS slurry (or 'juice', a one to ten mix of ABS scraps and acetone), hairspray and Pritt sticks applied to the print bed. There are even companies selling replacement print services as a third party addon.

PVA is very useful for her owners with dual head configurations (dualstrusion) as one head can be used to print this as support material with the other head dedicated to ABS or PLA to print out the material for the intended object. After printing the PVA support material can be washed away with water. Printing in this way allows objects to be produced that would otherwise fail to print properly or cleanly due to issues with overhangs yet still maintain a high quality finish.

What to do when it all goes wrong?

It can go all horribly wrong: The extruder is jammed. There is smoke coming out of machine. Your print stopped four fifths of its way through a ten-hour print. Someone deviously crossed the filament over on your particular reel and the filament feed is now jammed up. The print isn't sticking to the bed. Your printer is making a horrible grinding noise. Your printer has just stopped mid print for no readily apparent reason. The wonderful super bargain priced two kilo reel you bought has filament in the middle with a diameter well below spec and your extruder is now milling air. Your printer has cleverly decided to knock over the build of the fine vase that you were so looking forward to. All these things have happened to me.

There is no need to sit in a huddle and weep, although you might feel better for it.

3D printers are obtuse creatures and rarely warn you what has gone wrong or what is creating problems. You have to become a bit of a detective to think through the problems and find clues for the causes of issues you are facing. Patience and perseverance are called for here. I know I don't have all the answers to all my problems with 3D printing but I do know other can people short-circuit solving problems radically. When you are stuck the answer is often to ask for help.

Asking for help.

When you have a problem with your printer by all means raise a call with the manufacturer through their website or support forum. But do not sit back and think the job is done. Make sure you also check out the online community groups and maybe post a question to ask you some help. When you do post a question do make sure you don't resort to overemotional language. It doesn't help other people to have your emotions machine-gunned all over them. They are not the problem. They have their own emotions to deal with as well. So treat your problem like a scientific experiment; first method, then result and conclusion. Be concise and clear about what has gone wrong and in what circumstances it's gone wrong.

You need to ask yourself some questions. Is the problem repeatable what's the impact of the problem what else might have been much affected the situation. You should take great care that you are posting in the right area of the forum as otherwise this can greatly annoy those from whom you're seeking help. It makes sense to search beforehand case someone has already come across exactly the same problem something very close to and is posted the question was already received an answer. This way you can avoid the problem of people having to answer the same questions over and over again so read the FAQ sections of the forum first.

Some Answers

Lets answer some of the common generic questions and see what might help. I will answer some of the question posed above to help you.

If your extruder is jammed try posing the print have not already done so and try reversing the filament through the control panel. If this fails you may need to strip down the extruder and take apart and manually clear filament path with pliers or similar. Sometimes I have used a small hex spanner to push through some resistant filament but this needs to be done with great care in case you damage any of the extruder's linings.

In severe cases you may need to undo the nozzle of the printer (according to manufacturers instructions) and in the case of an ABS filament blockage soak in acetone. Care needs to be taken when doing this because of the fumes and obviously needs to be done when nozzle is cold, although you will probably need to take the nozzle off while it's hot first while the filament inside is soft and molten. You will need a glass jar with a lid and pour enough pure acetone to cover the nozzle and leave overnight. Avoid using nail polish acetone as it often comes with oils added which significantly reduces the effectiveness of the acetone. For PLA soak in a water and caustic soda mix. As an extreme last resort some people have used a very fine drill or even a blowtorch to burn off the print material but neither of these tactics is recommended. Be prepared to take your time, not to rush things and proceed with great caution in terms of safety. In case of doubt and you are not sure about doing any of this I would recommend replacing the blocked parts instead or finding the nearest 3D print or maker help group for assistance.

Sometimes when your print is messed up the only thing is to try and do the print again. Consider using a different batch or type of filament if you have extrusion problems. Adjustment of the extruder tolerances is another option too, keep a record of the changes you make so you can work through this methodically.

When you find your print fall over during printing there are a number of things you can do to fix this. One is switch on rafting when you slice your design so there is a bigger surface area for the design to stick to the print bed. You may need to switch on supports as well if the design is unstable (quite often when attempting to print someone who has been 3D scanned). If this is not enough you may need to look at improving the adhesion to the print bed by using ABS juice or dilute PVA. A switch over to a different feedstock, e.g. from ABS to PLA may help too. Lastly consider using a different type of 3D Printer altogether or send the print out to a bureau and see what results they can get for you.

If your printer is making odd noises you can see if you can pinpoint the part of the printer making the noise and make a note of which actions are triggering it. Is in happening when your printer is moving in only one axis? If so it narrows down the possibilities where problems may be occurring. Your printer will suffer from wear and tear to so odd noises may be an early warning that parts need to be replaced or repaired.

On the topic of safety, if your printer does start smoking switch it off straight away! I had to do this myself when the fuses blew on the heater board circuits of one of my printers. While It was possible to repair the failed board I chose to replace it instead as it was a relatively cheap item.

Conclusion: From Thought to Form.

T his book has led you, gently I hope, through a wide range of many of the aspects involved with 3D printing and the revolutionary effects it is starting to have upon our all lives.

We have explored what 3D printing actually is and realised something of what it's like to experience the sheer excitement and joy the printing process offers us.

From the Byzantine technical aspects we realise that although the technology can be intimidating at first appearance many of the mechanical aspects of 3D printing are relatively straightforward and easy to grasp. We know the difference between traditional subtractive manufacturing processes which require the removal of material to leave a desired shape as opposed to the additive methods used in 3D printing whereby objects are created one layer atop another, like sliced bread, to achieve the intended shape. We have seen that whether it is a basic machine or an industrial one the same principle holds for all, whatever the technical variations and diverse types of feedstock used to achieve the printing process.

We covered the processes which are required to take designs shapes, slice them then print them out. The different techniques of laser printing, FDM, binding and sintering are now familiar to us and we have a greater insight may help us decide which method may be more most appropriate to use to suit our particular needs.

The history of 3D printing has also been brought to light and we can now appreciate convolutions caused by the way in which technology has been developed subsequently sprung to life following the expiry of several key patents which kept us wonderful technology corralled in the backwaters of rapid prototyping. The parallels with the early computer industry become apparent and that analogy enlightens us to help foresee the possible directions in which this wonderful technology will develop in the future.

The synergies of 3D printing and the Internet were also highlighted as we may recall from the explosion of interest generated by the Reprap project. We can see robotics and 3D printing marching side-by-side into the future as another pair of complimentary technical developments grow together to intertwine and overlap as time passes.

Conclusion: From Thought to Form.

The economic impacts and business implications also been touched on so we have a clear understanding of the impact that will be felt in these areas. While it is evident that 3D printing will not supplant traditional manufacturing processes in the short to medium term it is likely given the growing diversification of materials and improvements in the print processes themselves that 3D printing will play a more significant role in many future manufacturing applications.

Designing in 3D and the tools you need to do this have also been covered and you will hopefully find such platforms readily accessible and should empower you to experiment and develop your ideas easily in such environments. It is possible to use many of these tools for little or no cost at all.

The variety of materials has grown dramatically and is a reflection of the increasing demand in the marketplace which is hungry for evermore combinations and permutations of feedstock intended specifically for the 3D printing market. We know the difference between a ABS and PLA filaments, the upcoming hybrids, printing in wood and even concrete.

Our knowledge of potential for good and the inherent risks for bad in the application of copyright laws for 3D printing designs and objects will help us skirt around the legal niceties without risking infringement. We can also appreciate the need to decide in what way we may share or charge for our designs before we risk uploading them on the Internet.

The power of the growing communities and the concept of Makers has been touched on we realise there are plenty of opportunities for us to get involved whatever level that suits us and enjoy participating with like-minded fellows the marvels this technology can grant us.

I hope you enjoyed the interlude of the story about the future of 3D printing and indeed our possible future as well. The ever blossoming applications are becoming clearer and we can see in the future 3D printing will be used for many more things, from serious medical applications, through custom wearable items and clothing, buildings and even towards construction in space.

We have looked at the growing number of different 3D printer designs and core architectures so we know what to look for should we choose to buy or build a 3D printer of our very own. We will not be afraid to ask for help and we needed and will know where to get it and how to ask for it. I'm sure many of us will be tempted to join in and enjoy the exhilaration of printing objects of our own design, from the ideas in our heads, to the designs in our computers, and finally the production of our own copies that very same design which we had originally conceived or modified, with permission, those of others.

Don't forget that additional material is also available from the website 3dprintingunleashednow.com free to all readers. You may be pleasantly surprised!

3D Printing Unleashed

I hope you have enjoyed this journey into the exciting world of 3D printing and I wish you all now feel empowered to embrace the opportunities this technology offers us all.

Appendix

7 Key Questions Answered Here

What is 3D printing?

3D printing is a way of making physical objects under computer control by combining and bonding feedstock material together. It is a form of additive manufacturing[143] used to generate an accurately formed three dimensional object. This is done in such a precise way that the process can be repeated to give almost exactly the same result time and time again. See chapter 1.

How does it work?

How 3D printing works is very simple. It puts one thing very precisely on top of another and bonds them together in some way. In other words it is an additive process. See Chapter 2.

How did it come about?

It derived from Computer Numerical Control (CNC) and rapid prototyping technologies in the 1980s. The invention of stereo lithography apparatus (SLA) by Charles Hull in 1983 may be considered the defining moment when the idea of 3D printing crystallised. See Chapter 3.

What can I make with it?

There is a multitude of different objects you can make, the list is growing as you read this. It may be tempting to dismiss 3D printing as only good for models or plastic toys but they have also been used in industry to print parts for Airbus aircraft. The capabilities of the 3D printers available are improving all the time. See chapter 6.

What do I need to look for when buying a 3D printer?

To answer this question you need to know what you want to use a 3D printer for. If you want to try one out then it makes little sense to go for a big industrial class printer. If it is for the home then you need a small lightweight printer. If you need fine detail consider a resin printing printer. If cost is your main concern an FDM extruder based model is a good choice. See chapter 9.

What's in it for business?

It can reduce lead times, save money and grant greater flexibility in responding to changing consumer needs and demands. While it is less likely to be cost effective for large production runs it can enhance the design stage of product development and related business decision making. See chapter 4.

Can I design for 3d printing?

Using CAD software or 3D design website platforms it is easy to get started in 3D print designing. Some of these are free to use too. There is a lot of existing designs on online repositories to give you an idea of the possibilities too. You can even get a test print of your design made without owning a 3D printer, by taking advantage of 3D printing bureau available worldwide. See chapter 5.

References

1) What is the definition of 3D printing? There are a variety to choose from and I do cover this in the book - wikipedia always has something to say http://en.wikipedia.org/wiki/3D_printing
2) http://www.eos.info/additive_manufacturing/for_technology_interested
3) See http://www.3dprinterprices.net/advantages-of-3d-printing-over-traditional-manufacturing-2/
4) https://en.wikipedia.org/wiki/Machining
5) See a comparison here http://www.approto.com/Media-Center/Additive-vs-Subtractive-Manufacturing--Which-is-Ri.aspx
6) They are great fun – see http://www.makerfaireuk.com/
7) A classic often used for test print of SLS printers, see the many examples of it here https://www.thingiverse.com/tag:Eiffel_Tower
8) The different varieties and designs for these are mindboggling. http://en.wikipedia.org/wiki/Quadcopter - There are self build options and even an arduino based concept - http://ardupilot.com/
9) One of the most successful is the 3d Printshow - http://3dprintshow.com/ They started off in 2012 and the list of venues keeps growing.
10) A Makerbot Thing-O-Matic. An open source wooden frame design which despite losing the odd nut and bolt it still works fine. http://www.makerbot.com/blog/2010/09/25/announcing-makerbots-new-3d-printer-the-thing-o-matic/
11) See https://en.wikipedia.org/wiki/Dry_stone
12) https://en.wikipedia.org/wiki/Raw_material
13) A most amazing statue, makes me feel it's about to draw breath whenever I have seen the original. http://www.accademia.org/explore-museum/artworks/michelangelos-david/
14) https://en.wikipedia.org/wiki/Voronoi_diagram
15) Some cool stuff here https://www.thingiverse.com/tag:Voronoi
16) Think of large Lego bricks piled one on top of each other … http://en.wikipedia.org/wiki/Ziggurat
17) http://en.wikipedia.org/wiki/Pyramid
18) https://en.wikipedia.org/wiki/Phase_transition
19) https://en.wikipedia.org/wiki/Laser_cutting
20) Both FDM and FFF are mentioned here https://en.wikipedia.org/wiki/Fused_deposition_modeling
21) The term adopted for the Reprap project to avoid any copyright infringement https://en.wikipedia.org/wiki/RepRap_Project
22) The company's website is http://www.stratasys.com/ also associated with the Objet brand after a merger with a company of that name in 2012
23) https://en.wikipedia.org/wiki/Thermoplastic

24) A respected and reliable design http://reprap.org/wiki/Wade%27s_Geared_Extruder

25) Ideal where a low print head weight and inertia are paramount http://reprap.org/wiki/Erik%27s_Bowden_Extruder

26) An exception to this was the early Makerbot Thing-O-Matic's cartesian design which moved the print bed in the X/Y directions (left and right) and the nozzle in the z direction (up and down).

27) http://3dprint.com/2140/topolabs-fdm-3d-print-method/

28) Like ABS http://www.reprap.org/wiki/ABS

29) https://en.wikipedia.org/wiki/Selective_laser_sintering

30) More materials science https://en.wikipedia.org/wiki/Sintering

31) https://en.wikipedia.org/wiki/Selective_laser_melting

32) CNC, an acronym with several meanings, but what is meant here can found at https://en.wikipedia.org/wiki/Numerical_control

33) Some patents are still waiting in the wings http://www.3ders.org/articles/20140128-let-the-revolution-begin-key-3d-printing-patent-expires-today.html

34) https://en.wikipedia.org/wiki/Computer-aided_design

35) Now it all seems ancient history but the PC wasn't originally considered a dead certainty https://en.wikipedia.org/wiki/IBM_Personal_Computer

36) https://en.wikipedia.org/wiki/Wire-frame_model

37) https://en.wikipedia.org/wiki/Component_parts_of_internal_combustion_engines

38) https://en.wikipedia.org/wiki/Clay_modeling

39) https://en.wikipedia.org/wiki/Rapid_prototyping

40) http://3dprintingindustry.com/3d-printing-basics-free-beginners-guide/history/

41) https://en.wikipedia.org/wiki/Stereolithography

42) https://en.wikipedia.org/wiki/Chuck_Hull

43) https://en.wikipedia.org/wiki/3D_Systems

44) http://www.printbotz.com/what-is-3d-printing

45) https://en.wikipedia.org/wiki/Stratasys

46) http://www.eos.info/

47) There is no one definitive history of 3D printing as yet, some good yet flawed attempts have been made e.g. http://www.engineering.com/3DPrinting/3DPrintingArticles/ArticleID/6262/Infographic-The-History-of-3D-Printing.aspx

48) One perspective https://www.techdirt.com/articles/20120130/16535017591/how-patents-have-held-back-3d-printing.shtml and a contrasting view can be found here http://www.livescience.com/38494-3d-printing-and-patent-protection.html with open source retro-active patenting ongoing to address patent trolling https://www.eff.org/deeplinks/2013/03/effs-fight-open-3d-printing-contin-

ues-askpatentscom

49) A brief history here too https://en.wikipedia.org/wiki/RepRap_Project

50) As applied in software https://en.wikipedia.org/wiki/Open-source_software historical information here https://en.wikipedia.org/wiki/Open_Source_Initiative

51) Plenty of development ideas for this hardware platform evident http://www.arduino.cc/

52) http://www.tridprinting.com/3D-Printing-Patents/ with a reflection on what it covers here http://payne.org/blog/stratasys-heated-build-enclosure-patent/

53) https://en.wikipedia.org/wiki/Maker_culture

54) A form of online digital library see https://en.wikipedia.org/wiki/Digital_library

55) https://en.wikipedia.org/wiki/G-code

56) Will be interesting to see if this will be in common use in the future http://inthefold.autodesk.com/in_the_fold/2014/05/accelerating-the-future-of-3d-printing.html

57) https://en.wikipedia.org/wiki/Gold_rush

58) One rapidly growing example are 3d hubs, who claimed at the 2015 3D printshow to operate in more countries than McDonalds. https://www.3dhubs.com/

59) Once a big name in the field http://www.computernostalgia.net/articles/sinclairResearch.htm

60) https://en.wikipedia.org/wiki/Microwriter

61) An initial announcement http://www.makerbot.com/blog/2013/06/19/makerbot-and-stratasys-announce-merger/ and news of recent rationalisation http://richrap.blogspot.co.uk/2015/04/handling-things-makerbot-way.html

62) A balanced article on this can be found here http://www.shapeways.com/blog/archives/1933-Comparing-Apples-and-Oranges-Injection-Molding-vs-3D-Printing.html

63) According to this article in the bbc news in the 'How Lego was built ' section 'usually about $50,000 (£32,000)' see http://www.bbc.co.uk/news/uk-politics-29992974

64) For the archive see http://www.bbc.co.uk/archive/tomorrowsworld/

65) See http://www.dezeen.com/2012/11/12/us-military-invests-in-3d-printing-on-the-frontline/

66) The RepRap project recommends a supply of about 240w http://reprap.org/wiki/Choosing_a_Power_Supply_for_your_RepRap

67) My Leapfrog Creatr with a heated bed has a powers supply rated at 400w http://www.lpfrg.com/creatr-power-supply

68) See future 3D business trends here http://3dprintingindustry.com/2015/05/13/majority-businesses-plan-increase-3d-printing-spending-sculpteo-study-reveals/

69) Not just tools they are also thinking of food too http://www.nasa.gov/directorates/spacetech/home/feature_3d_food_prt.htm

70) A useful article to help with this is on the Forbes website http://www.forbes.com/sites/rebeccabagley/2014/05/03/how-3d-printing-can-transform-your-business/ and an earlier article on the entrepreneur website is also informative http://www.entrepreneur.com/article/225446

71) It is a grey area – see http://www.3ders.org/articles/20150122-man-accused-of-copyright-infringement-after-3d-scanning-a-michelangelo-statue.html and http://newmediarights.org/legal_issues_arise_creating_3d_file_scanning_object_0 for starters

72) See here for details http://www.david-3d.com/en/

73) https://en.wikipedia.org/wiki/Computer-aided_manufacturing

74) https://en.wikipedia.org/wiki/G-code

75) https://en.wikipedia.org/wiki/Open-source_software

76) Openscad's homepage is http://www.openscad.org/

77) See https://en.wikibooks.org/wiki/OpenSCAD_User_Manual

78) See http://www.blender.org/

79) See https://inkscape.org/en/

80) Vector graphics is covered here https://en.wikipedia.org/wiki/Vector_graphics

81) An example can seen in MichaelAtOz's Lithophane on thingiverse http://www.thingiverse.com/thing:78719

82) See http://meshlab.sourceforge.net/

83) Another form of open source style licensing se https://en.wikipedia.org/wiki/GNU_General_Public_License

84) Explained here https://en.wikipedia.org/wiki/Proprietary_software

85) AutoCAD by Autodesk main site is http://www.autodesk.com/

86) More at https://en.wikipedia.org/wiki/Autodesk

87) Covered here https://en.wikipedia.org/wiki/.dwg

88) The Scottish software is found here http://www.anarkik3d.co.uk/

89) Now from here http://www.sketchup.com/ for company history see https://en.wikipedia.org/wiki/SketchUp

90) Available from http://www.rhino3d.com/

91) B-splines see https://en.wikipedia.org/wiki/Non-uniform_rational_B-spline

92) Excellent community based 3d printing bureau https://www.3dhubs.com/

93) See https://www.tinkercad.com/

94) Grab it from here http://slic3r.org/

95) See http://www.kisslicer.com/

96) Available from the successful 3D print manufacturer Ultimaker http://wiki.ultimaker.com/Cura

97) http://reprap.org/wiki/Erik%27s_Bowden_Extruder

98) Background here http://www.reprap.org/wiki/Skeinforge and in more detail here http://fabmetheus.crsndoo.com/wiki/index.php/Skeinforge

99) Available from http://www.netfabb.com/

100) See http://apps.123dapp.com/3dprint/

101) See http://www.123dapp.com/meshmixer

102) Blogged here http://blog.123dapp.com/2013/07/introducing-the-autodesk-3d-print-utility

103) The final leg of the print process to control the printer and feeed it its instructions see https://en.wikipedia.org/wiki/Computer-aided_manufacturing

104) Sources are http://reprap.org/wiki/ReplicatorG or http://www.sailfish-firmware.com

105) Extremely good firmware for Makerbot and related printers http://www.-sailfishfirmware.com

106) Available here http://www.pronterface.com/

107) Available at http://www.repetier.com

108) See http://www.repetier-apps.com/ for details.

109) Available from https://www.simplify3d.com/software/

110) Such as http://www.lpfrg.com/store/3d-printer/software/simplify3d-software

111) ItsJustMidnight - in a reply to Artbot http://www.thingiverse.com/make:100383

112) See http://www.filabot.com/

113) In their own words http://www.filabot.com/pages/about-us

114) The cubify printers being a case in point, see their site http://cubify.com/, and also for Stratsys Objets printers as well, e.g. http://www.tritech3d.co.uk/objet-3d-printer-cartridges.php

115) From the UK http://www.faberdashery.co.uk/

116) For reference https://creativecommons.org/

117) Implications on the ethics of this discussed here http://www.washington-post.com/national/health-science/smart-pills-with-chips-cameras-and-robotic-parts-raise-legal-ethical-questions/2014/05/24/6f6d715e-dabb-11e3-b745-87d39690c5c0_story.html

118) More here http://www.thinkgig.com/the-internet-of-things-and-healthcare/ and here https://www.salesforce.com/blog/2014/05/hospitals-internet-of-things-patient-care.html

119) A good read, https://en.wikipedia.org/wiki/Nineteen_Eighty-Four

120) Found here http://www.thingiverse.com/

121) Check the small print especially section 3.2 http://www.thingiverse.com/legal

122) See https://grabcad.com/

123) Well worth checking out here http://www.shapeways.com

124) Another one for the bookmark https://pinshape.com

125) For a comparison between Thingiverse and Pinshape see http://3dprintingindustry.com/2015/06/17/pinshape-vs-thingiverse-showdown-in-3d-model-town/

126) Available at http://i.materialise.com/

127) Here they are again http://www.shapeways.com

128) See https://www.3dhubs.com/

129) http://www.3dcreationlab.co.uk/

130) Siemens Vice President Andreas Saar discusses this: http://3dprintingindustry.com/2015/05/21/siemens-vp-andreas-saar-on-all-things-hybrid-3d-printing/

131) One example of the advantage a 3D printed model gave surgeons can be seen here http://www.bbc.co.uk/news/health-30996506

132) See http://www.nasa.gov/content/open-for-business-3-d-printer-creates-first-object-in-space-on-international-space-station

133) http://www.makerbot.com/

134) See informed comment here on the subject http://richrap.blogspot.co.uk/2014/05/makerbot-patents-twist-knife-on-open.html

135) https://ultimaker.com/

136) http://www.stratasys.com/

137) http://www.lpfrg.com/

138) http://the3doodler.com/

139) https://www.kickstarter.com/

140) http://the3doodler.com/

141) Here's that link again https://www.kickstarter.com/

142) Excellent solution for remote host printing http://octoprint.org/

143) http://www.eos.info/additive_manufacturing/for_technology_interested

Glossary

3D Printing

A form of additive manufacturing where objects are created under computer control from a 3D design model.

ABS

Acrylonitrile Butadiene Styrene is the thermoplastic Lego is made from. A good material to start with as it will print quite well even if the temperature (220-240°C) is not set quite right. A small amount mixed with pure acetone (about 1:10) can be used to make 'ABS juice' to help prints stick to the print bed. Tends to curl on the edges of a print at times which can be a nuisance.

Additive Manufacturing

A means of building three dimensional objects by depositing or extruding layers and fusing them together. Sometimes used interchangeably with the term 3D printing.

CAD

Computer Aided Design. Design using computer software to create, copy, modify or combine 3D designs as virtual models. Often allows the 3D design model to be displayed as a 3D computer graphic.

Extruder

The part of the printer which pushes the feedstock towards the heat block and of the print nozzle (the hot end). Often powered by a stepper motor.

FDM/FFF

Fused deposition modeling (trademarked by Stratasys Inc) or fused filament fabrication, a form of 3D printing using a filament feedstock which is heated, extruded into layers and fused together before cooling.

Filament

Feedstock material used to make printed objects for FDM/FFF printers. It is often offered in either 1.75 mm or 3mm diameters.

Firmware

The embedded control software that runs on the computer electronics of the 3D printer.

Gcode

The detailed set of instructions a 3D printer understands which tells it how to follow the correct print path and make the desired print object. Usually generated for you by slicing software.

Guide Hole

Where the filament goes in to load the extruder with feedstock. The hot end may need to be warmed up first depending upon the printer type.

Hot End

The heat block and nozzle used in the FDM/FFF process to heat and convert the feedstock into a printable form.

Infill

The interior of a printed object. Its structure and density is often determined by the settings used in the slicing process. A solid infill is where a 100% infill setting is used so that the printed object has no gaps or voids inside.

Kapton Tape

Polyamide adhesive tape which can withstand temperatures of up to 250°C. This and strong with a yellow colour and a tendency to tear easily. Sometimes used to improve print bed or platform surfaces

Print Bed or Build Platform

The area within a print where a print design starts to be printed. Usually a flat glass or metal area and may be heated if the feedstock requires it (usually for ABS).

Raft

An extra layer or two of material printed first to provide a sturdier base for a tricky print that has a small surface area that would otherwise be in contact with the print bed. There is often an option to enable this in the slicing software being used.

Repository

A library to store and organise print designs, often online.

Shell

a value used to determine the wall thickness or skin of a print object. Usually a factor of the nozzle diameter.

Slicing

The process of converting a 3D design into virtual layers that can be used as to generate print paths for a 3D printer to follow to generate a print. Output is often in the form of Gcode which is specific to the type of 3D printer intended to be used.

Stepper Motor

A motor which translates electric pulses into rotational segments, thereby simplifying electronic control of the rotational motion produced.

Stereolithography

An additive manufacturing process which works by selectively solidifying layers of photo sensitive resin. This is usually achieved with the beams from ultraviolet (UV) lasers. The printed object may need additional curing after printing to achieve full strength and stability.

Support Material or Support Structures

Additional print material generated to support areas of a print that would otherwise deform during printing. Differing feedstock materials may be used to make removal after printing easier. This is great to use to compensate for difficult designs, or to print 3D scans of something, where the design constraints of a 3D printer were not considered. There is often an option to enable support structures in the slicing software being used.

STL File

Intermediate file in a 3D printing process often created after a design has been created but before it has been sliced to produce Gcode. It describes the surface of an object in terms of triangular coordinates.

Subtractive Manufacturing

The traditional means of making desired objects by the removal of parts of a stock material to reveal the desired shape. The removed material is often treated as waste although sometimes it can be recycled. Techniques such as turning, cutting, milling, routing, boring or drilling material is removed to generate the desired object by removing some of it from a stock material.

About the Author

A man with a passion for technology and ingenuity. Peter Goodwin was enthralled in watching computer business flourish from first saving up for a home computer and making many modifications to it in the 1980's to go on to work in this fascinating field ever since.

The up swell of interest in 3D printing, typified by a special report in NewScientist magazine in July 2011, drew his attention towards 3D printing

He built his first printer a dual extrusion Makerbot Thing-o-Matic shortly thereafter and recently put together a ReprapPro beta version Fisher Delta.

He is an avid attendee of every London 3D Printshow since its inception in 2012 and backed the successful 3Doodler kickstarter project of 2013.